Encyclopaedia of

World Geography

Encyclopaedia of
World Geography

Vol. - 2

Chandra Vijay Purty

ABD PUBLISHERS
Jaipur ■ New Delhi

ABD PUBLISHERS

Regd. Off. "Bony Residency", Gate No. 2
Opp. Tilak Public School, Vishweswariya Nagar,
Gopalpura Road, Jaipur - 302018, India
Ph: 0141 - 2761280, Telefax: 2761381
Email: abdpublishersjpr@gmail.com

Delhi Office: 102, 1st Floor, Satyam House,
4327/3, Ansari Road, Darya Ganj,
New Delhi - 110002, India
Ph: 011-45652440

visit us: www.oxfordbookcompany.com

Published in India, 2024

ISBN : 9788183763127 (7 Vols. Set)

Typeset by Shivangi Computer, Jaipur

Printed and bound in India by Thomson Press Ltd.,
New Delhi

Preface

Geography is the study of the Earth and its lands, features, inhabitants, and phenomena. A literal translation would be "to describe or write about the Earth". The first person to use the word "geography" was Eratosthenes (276-194 B.C.). Four historical traditions in geographical research are the spatial analysis of natural and human phenomena (geography as a study of distribution), area studies (places and regions), study of man-land relationship, and research in earth sciences. Nonetheless, modern geography is an all-encompassing discipline that foremost seeks to understand the Earth and all of its human and natural complexities—not merely where objects are, but how they have changed and come to be. As "the bridge between the human and physical sciences," geography is divided into two main branches—human geography and physical geography.

The oldest known world maps date back to ancient Babylon from the 9th century BC. The best known Babylonian world map, however, is the *Imago Mundi* of 600 BC. The map as reconstructed by Eckhard Unger shows Babylon on the Euphrates, surrounded by a circular landmass showing Assyria, Urartu and several cities, in turn surrounded by a "bitter river" (Oceanus), with seven islands arranged around it so as to form a seven-pointed star. The accompanying text mentions seven outer regions beyond the encircling ocean. The descriptions of five of them have survived. In contrast to the *Imago Mundi*, an earlier Babylonian world map dating back to the 9th century BC depicted Babylon as being further north from the centre of the world, though it is not certain what that centre was supposed to represent.

The ideas of Anaximander (c. 610 B.C.-c. 545 B.C.), considered by later Greek writers to be the true founder of geography, come to us through fragments quoted by his successors. Anaximander is credited with the invention of the gnomon, the simple yet efficient Greek instrument that allowed the early measurement of latitude. Thales, Anaximander is also credited with the prediction of eclipses. The foundations of geography can be traced to the ancient cultures, such as the ancient, medieval, and early modern Chinese. The Greeks, who were the first to explore geography as both art and science, achieved this through Cartography, Philosophy, and Literature, or through Mathematics. There is some debate about who was the first person to assert that the Earth is spherical in shape, with the credit going either to Parmenides or Pythagoras. Anaxagoras was able to demonstrate that the profile of the Earth was circular by explaining eclipses. However, he still believed that the Earth was a flat disk, as did many of his contemporaries. One of the first estimates of the radius of the Earth was made by Eratosthenes.

The first rigorous system of latitude and longitude lines is credited to Hipparchus. He employed a sexagesimal system that was derived from Babylonian mathematics. The parallels and meridians were sub-divided into 360°, with each degree further subdivided 602 (minutes). To measure the longitude at different location on Earth, he suggested using eclipses to determine the relative difference in time. The extensive mapping by the Romans as they explored new lands would later provide a high level of information for Ptolemy to construct detailed atlases. He extended the work of Hipparchus, using a grid system on his maps and adopting a length of 56.5 miles for a degree.

From the 3rd century onwards, Chinese methods of geographical study and writing of geographical literature became much more complex than what was found in Europe at the time (until the 13th century). Chinese geographers such as Liu An, Pei Xiu, Jia Dan, Shen Kuo, Fan Chengda, Zhou Daguan, and Xu Xiake wrote important treatises, yet by the 17th century, advanced ideas and methods of Western-style geography were adopted in China.

This book contains the fundamental and basic information of the subject.

Chandra Vijay Purty

Contents

1

Australia: A Settler Society in a Changing World

Before the arrival of European settlers, Aboriginal and Torres Strait Islander peoples inhabited most areas of the Australian continent. They spoke one or more of hundreds of separate languages and dialects, and their lifestyles and cultural traditions differed from region to region. Their complex social systems and highly developed traditions reflect a deep connection with the land.

Asian and Oceanic mariners and traders were in contact with Indigenous Australians for many centuries before the era of European expansion. Some formed substantial relationships with communities in northern Australia.

European Settlement

The first recorded European contact with Australia was in March 1606, when Dutch explorer Willem Janszoon (1571–1638) charted the west coast of Cape York Peninsula in Queensland. Later that year, the Spanish explorer Luis Vaez de Torres sailed through the strait separating Australia and Papua New Guinea. Over the next two centuries, European explorers and traders continued to chart the coastline of Australia, then known as New Holland. In 1688, William Dampier became the first British explorer to land on the Australian north west coast. It was not until 1770 that another Englishman, Captain James Cook, aboard the Endeavour, extended a scientific voyage to the South Pacific in order to further chart the east coast of Australia and claim it for the

British Crown. Britain decided to use its new outpost as a penal colony. The First Fleet of 11 ships carried about 1500 people – half of them convicts. The fleet arrived in Sydney Harbour on 26 January 1788, and it is on this day every year that Australia Day is celebrated.

About 160 000 men and women were brought to Australia as convicts from 1788 until penal transportation ended in 1868. The convicts were joined by free immigrants beginning in the early 1790s. The wool industry and the gold rushes of the 1850s provided an impetus for increasing numbers of free settlers to come to Australia.

Scarcity of labour, the vastness of the land and new wealth based on farming, mining and trade made Australia a land of opportunity. Yet during this period Indigenous Australians suffered enormously. Death, illness, displacement and dispossession disrupted traditional lifestyles and practices.

A Nation is Born

The Commonwealth of Australia was formed in 1901 through the federation of six states under a single constitution. The non-Indigenous population at the time of Federation was 3.8 million, while the estimated Indigenous population was around 93,000. Half of the people lived in cities, three-quarters were born in Australia, and the majority were of English, Scottish or Irish descent.

The founders of the new nation believed they were creating something new and were concerned to avoid the pitfalls of the old world. They wanted Australia to be harmonious, united and egalitarian, and had progressive ideas about human rights, the observance of democratic procedures and the value of a secret ballot.

One of the first acts of the new Commonwealth Parliament was to pass the Immigration Restriction Act 1901, which limited migration to people of primarily European origin. This was dismantled progressively after the Second World War. Today Australia has a global, non-discriminatory policy and is home to people from more than 200 countries.

From 1900 to 1914 great progress was made in developing Australia's agricultural and manufacturing capacities, and in setting up institutions for government and social services.

The Impact of War

In the end ANZAC [Australian and New Zealand Army Corps] stood and still stands for reckless valour in a good cause, for enterprise, resourcefulness, fidelity, comradeship and endurance that will never admit defeat.

Charles Bean, Official Historian of the First World War

The First World War had a devastating effect on Australia. In 1914 the male population of Australia was less than 3 million, yet almost 400,000 of them volunteered to fight in the war. An estimated 60,000 died and tens of thousands were wounded.

The period between the two world wars was marked by instability. Social and economic divisions widened during the Depression years when many Australian financial institutions failed.

During the Second World War, Australian forces made a significant contribution to the allied victory in Europe, Asia and the Pacific. The generation that fought in the war and survived came out of it with a sense of pride in Australia's capabilities.

ANZAC DAY by Michael McKernan

'Excuse him, sir', an Australian soldier said of his drunken companion to an officer in northern France in 1917, the war's second last year, 'he's been keeping up Anzac'. Soldiers in France had special reason to remember 25 April 1915. At dawn on that day they, or mates with whom they were now serving, landed at Gallipoli in Turkey, and within days had won a reputation for courage and endurance that all Australia celebrated.

From the outset remembrance of that special Australian day involved the quiet of church and personal reflection, and a coming together of former comrades to recall the horror of war, the memory of lost mates, and a joyful celebration that some were still alive. They remembered with pride all that the Australians had achieved.

The rituals of the day emerged through the 1920s and were reinterpreted and reinvigorated in the 1990s. The Anzacs had landed at dawn and it seemed appropriate that they be commemorated at exactly that time of day. Originally restricted to returned soldiers only, the religious service held at dawn on 25 April each year was short but moving, recalling particularly the first morning on Gallipoli above other Australian actions in war. But, by extension and over time, all Australians who have lost or risked their lives in war are now remembered.

By the late 1920s men who had served together in war came together in reunions after the dawn service, to live again the most adventurous and most terrible times of their lives, and to march together in unity and fellowship in the major cities and smaller towns – to march, as war historian Charles Bean had written, 'down the long lane of [their] country's history'. The Anzac Day march peaked in numbers in the 1950s when the men and women of the Second World War joined the original Anzacs. In Melbourne and Sydney several city blocks would be filled with marchers, in their original battle groupings but in civilian clothing, applauded with delight by tens of thousands of spectators, family and friends.

Yet there was some ambivalence about 'the Day' – too much drink, too much maudlin sentimentality, too much exclusion of those who had not fought. As the men of both wars aged, particularly the men who had fought in the First World War, the ambivalence dissipated and Anzac Day took on a new and deeper meaning. The marchers seemed so frail that the remembrance ceremonies began to take on a different character. They evoked great pride, family memories, deep respect. There were possibly more tears at the march in the 1990s than there had been in the earlier times.

Or was it that the tears had been hidden by more stoical generations? Jack Fothergill, a young man of 25 years from rural Victoria, landed at Anzac within a couple of hours of the attack. By breakfast time he was dead on Pine Ridge. For the next 30 years his parents remembered each Anzac Day with poems for the 'In Memoriam' newspaper notices that were different each year and deeply moving. 'If only I could see your grave, I would die happy', Jack's mother had written in 1923, her grief a permanent presence in her life. Australians now seem to understand that central fact of Anzac Day in the quiet reflection that the day still provokes.

Michael McKernan is a historian with a particular interest in Australia at war.

Post-war Prosperity

After 1945 Australia entered a boom period. Hundreds of thousands of refugees and migrants arrived in Australia in the immediate post-war period, many of them young people eager to embrace their new lives with energy and vigour. The number of Australians employed in the manufacturing industry had grown steadily since the beginning of the century. Many women who had taken over factory work while men were away at war were able to continue working in peacetime.

The economy developed strongly in the 1950s with major nation-building projects such as the Snowy Mountains Scheme, a hydro-electric power scheme located in Australia's south-east mountain region. Suburban Australia also prospered. The rate of home ownership rose dramatically from barely 40 per cent in 1947 to more than 70 per cent by the 1960s.

Other developments included the expansion of government social security programs and the arrival of television. Melbourne hosted the Olympic Games of 1956, shining the international spotlight on Australia. (In 2000, the Olympic Games came to Australia a second time, hosted by Sydney.)

A Changing Society

The 1960s was a period of change for Australia. The ethnic diversity produced by post-war immigration, the United Kingdom's increasing focus on Europe, and the Vietnam War (to which Australia sent troops) all contributed to an atmosphere of political, economic and social change.

In 1967 the Australian people voted overwhelmingly in a national referendum to give the federal government the power to pass legislation on behalf of indigenous Australians and to include indigenous Australians in future censuses. The referendum result was the culmination of a strong campaign by both indigenous and non-indigenous Australians. It was widely seen as a strong affirmation of the Australian people's wish to see their government take direct action to improve the living conditions of Aboriginal and Torres Strait Islander peoples.

The long post-war domination of national politics by the coalition of the Liberal and Country (now National) parties ended in 1972, when the Australian Labour Party was elected. The next three years saw major changes in Australia's social and economic policy agenda and extensive reforms in health, education, foreign affairs, social security and industrial relations. In 1975, however, the Governor-General dismissed the Labour Government, sparking a constitutional crisis. In the subsequent general election, the Labour Party suffered a major defeat and the Liberal-National Coalition ruled until 1983.

The Hawke-Keating Labour governments were in office from 1983 till 1996. They introduced a number of economic reforms, such as deregulating the banking system and floating the Australian dollar. In 1996 a Coalition Government led by John Howard won the general election and was re-elected in 1998, 2001 and 2004. The Liberal-National Coalition Government enacted several reforms, including changes in the taxation and industrial relations systems. In 2007 Mr. Kevin Rudd led the Australian Labour Party to government with policies designed to build a modern Australia equipped to meet the challenges of the future – including tackling climate change, reforming Australia's health and hospital system, investing in education and skills training and reforming Australia's workplace laws.

Regionalism

In politics, regionalism is a political ideology that focuses on the interests of a particular region or group of regions, whether traditional or formal (administrative divisions, country subdivisions, political divisions, subnational units). Regionalism centers on increasing the region's influence and political power, either through movements for limited form of autonomy (devolution, states' rights, decentralization) or through stronger measures for a greater degree of autonomy (sovereignty, separatism, independence). Regionalists often favour loose federations or confederations over a unitary state with a strong central government. Regionalism may be contrasted with *nationalism*.

Proponents of regionalism say that strengthening a region's governing bodies and political powers within a larger country would create efficiencies of scale to the region, promote decentralization, develop a more rational allocation of the region's resources for benefit of the local populations, increase the efficient implementation of local plans, raise competitiveness levels among the regions and ultimately the whole country, and save taxpayers money.

In some countries, the development of regionalist politics may be a prelude to further demands for greater autonomy or even full separation, especially when ethnic and cultural disparities are present. This was demonstrated in the late 1980s in Yugoslavia, among other examples.

A regionalist party is a regional political party promoting autonomy for its region; a regional party is a political party with its base almost entirely in a single region. All regionalist parties are also regional, while only a portion of regional parties are also regionalist. Because regional parties often cannot receive enough votes or legislative seats to be politically powerful, they may join political alliances or seek to be part of a coalition government.

Examples of regionalist parties include Regionalist Action Party of Chile in Chile (northern parts), the Savoy Region Movement in France (Savoy), The Friesen and South Schleswig Voter Federation in Germany (Friesland) and Telugu Desam Party in India (Andhra Pradesh). Examples of regional parties include the regionalist parties cited before, the Christian Social Union of Bavaria and almost all Belgian parties.

Regionalism may or may not include autonomism. Examples of autonomist parties include Action democratique du Quebec in Canada (Quebec), New Democratic Macau Association in China (Macau), Martinican Progressive Party (Martinique) and Communist Party of Reunion (Reunion) in France, Lega Nord in Italy (Northern Italy) and Popular Democratic Party of Puerto Rico in the United States (Puerto Rico).

One must distinguish between regionalist parties which support autonomy and nationalist/independentist movements which support independence. In the latter are included the likes of Frente para a Libertacao do Enclave de Cabinda (FLEC) in Angola (Cabinda province), the Nieuw-Vlaamse Alliantie and the Vlaams Belang and in Belgium (Flanders), the Basque Nationalist Party (Basque country), Convergence and Union

(Catalonia) and the Republican Left of Catalonia (Catalonia) in Spain and the Scottish National Party (Scotland) and Plaid Cymru (Wales) in the United Kingdom. Sometimes, the Lega Nord has formerly been independentist too.

Nineteenth Century Urbanization

Urbanization is the physical growth of urban areas as a result of global change. Urbanization is also defined by the United Nations as movement of people from rural to urban areas with population growth equating to urban migration. The United Nations projected that half of the world's population would live in urban areas at the end of 2008.

Urbanization is closely linked to modernization, industrialization, and the sociological process of rationalization.

Movement

As more and more people leave villages and farms to live in cities, urban growth results. The rapid growth of cities like Chicago in the late 19th century and Shanghai a century later can be attributed largely to rural-urban migration. This kind of growth is especially commonplace in developing countries.

The rapid urbanisation of the world's population over the twentieth century is described in the 2005 Revision of the UN World Urbanization Prospects report. The global proportion of urban population rose dramatically from 13% (220 million) in 1900, to 29% (732 million) in 1950, to 49% (3.2 billion) in 2005. The same report projected that the figure is likely to rise to 60% (4.9 billion) by 2030.. However, French economist Philippe Bocquier, writing in THE FUTURIST magazine, has calculated that "the proportion of the world population living in cities and towns in the year 2030 would be roughly 50%, substantially less than the 60% forecast by the United Nations (UN), because the messiness of rapid urbanization is unsustainable. Both Bocquier and the UN see more people flocking to cities, but Bocquier sees many of them likely to leave upon discovering that there's no work for them and no place to live."

According to the UN State of the World Population 2007 report, sometime in the middle of 2007, the majority of people worldwide will be living in towns or cities, for the first time in history; this is referred to as the arrival of the "Urban

Millennium" or the 'tipping point'. In regard to future trends, it is estimated 93% of urban growth will occur in developing nations, with 80% of urban growth occurring in Asia and Africa.

Urbanization rates vary between countries. The United States and United Kingdom have a far higher urbanization level than China, India, Swaziland or Niger, but a far slower annual urbanization rate, since much less of the population is living in a rural area.

- Urbanization in the United States never reached the Rocky Mountains in locations such as Jackson Hole, Wyoming; Telluride, Colorado; Taos, New Mexico; Douglas County, Colorado and Aspen, Colorado. The state of Vermont has also been affected, as has the coast of Florida, the Birmingham-Jefferson County, AL area, the Pacific Northwest and the barrier islands of North Carolina.
- In the United Kingdom, two major examples of new urbanization can be seen in Swindon, Wiltshire and Milton Keynes, Buckinghamshire. These two towns show some of the quickest growth rates in Europe.

Causes

Urbanization occurs naturally from individual and corporate efforts to reduce time and expense in commuting and transportation while improving opportunities for jobs, education, housing, and transportation. Living in cities permits individuals and families to take advantage of the opportunities of proximity, diversity, and marketplace competition.

People move into cities to seek economic opportunities. A major contributing factor is known as "rural flight". In rural areas, often on small family farms, it is difficult to improve one's standard of living beyond basic sustenance. Farm living is dependent on unpredictable environmental conditions, and in times of drought, flood or pestilence, survival becomes extremely problematic. In modern times, industrialization of agriculture has negatively affected the economy of small and middle-sized farms and strongly reduced the size of the rural labour market.

Cities, in contrast, are known to be places where money, services and wealth are centralized. Cities are where fortunes are made and where social mobility is possible. Businesses, which generate jobs and capital, are usually located in urban areas. Whether the source is trade or tourism, it is also through the cities that foreign money flows into a country. It is easy to see why someone living on a farm might wish to take their chance moving to the city and trying to make enough money to send back home to their struggling family.

There are better basic services as well as other specialist services that aren't found in rural areas. There are more job opportunities and a greater variety of jobs. Health is another major factor. People, especially the elderly are often forced to move to cities where there are doctors and hospitals that can cater for their health needs. Other factors include a greater variety of entertainment (restaurants, movie theaters, theme parks, etc) and a better quality of education, namely universities. Due to their high populations, urban areas can also have much more diverse social communities allowing others to find people like them when they might not be able to in rural areas.

These conditions are heightened during times of change from a pre-industrial society to an industrial one. It is at this time that many new commercial enterprises are made possible, thus creating new jobs in cities. It is also a result of industrialization that farms become more mechanized, putting many labourers out of work. This is currently occurring fastest in India.

Economic Effects

In recent years, urbanization of rural areas has increased. As agriculture, more traditional local services, and small-scale industry give way to modern industry the urban and related commerce with the city drawing on the resources of an ever-widening area for its own sustenance and goods to be traded or processed into manufactures.

Research in urban ecology finds that larger cities provide more specialized goods and services to the local market and surrounding areas, function as a transportation and wholesale hub for smaller places, and accumulate more capital, financial service provision, and an educated labour force, as well as often concentrating administrative functions for the area in which they lie. This relation among places of different sizes is called the urban hierarchy. As cities develop, effects can include a dramatic increase in costs, often pricing the local working class out of the market, including such functionaries as employees of the local municipalities. This is likely due the prevailing south-west wind which carries coal smoke and other airborne pollutants downwind, making the western edges of towns preferable to the eastern ones. Similar problems now affect the developing world, rising inequality resulting from rapid urbanisation trends. The drive for rapid urban growth and often efficiency can lead to less equitable urban development, think tanks such as the Overseas Development Institute have even proposed policies that encourage labour intensive growth as a means of absorbing the influx of low skilled and unskilled labour.

Urbanization is often viewed as a negative trend, but can in fact, be perceived simply as a naturally occurrence from individual and corporate efforts to reduce expense in commuting and transportation while improving opportunities for jobs, education, housing, and transportation. Living in cities permits individuals and families to take advantage of the opportunities of proximity, diversity, and marketplace competition.

Environmental Effects

The urban heat island has become a growing concern and is increasing over the years. The urban heat island is formed when industrial and urban areas are developed and heat becomes more abundant. In rural areas, a large part of the incoming solar energy is used to evaporate water from vegetation and soil. In cities, where less vegetation and exposed soil exists, the majority of the sun's energy is absorbed by urban structures and asphalt. Hence, during warm daylight hours, less evaporative cooling in cities allows surface temperatures to rise higher than in rural areas. Additional city heat is given off by vehicles and factories, as well as by industrial and domestic heating and cooling units. This effect causes the city to become 2 to 10° F (1 to 6° C) warmer than surrounding landscapes.. Impacts also include reducing soil moisture and intensification of carbon dioxide emissions.

In his book *Whole Earth Discipline*, Stewart Brand argues that the effects of urbanization are on the overall positive for the environment. Firstly, the birth rate of new urban dwellers falls immediately to replacement rate, and keeps falling. This can prevent overpopulation in the future. Secondly, it puts a stop to destructive subsistence farming techniques, like slash and burn agriculture. Finally, it minimizes land use by humans, leaving more for nature.

Changing form of Urbanization

Different forms of urbanization can be classified depending on the style of architecture and planning methods as well as historic growth of areas.

In cities of the developed world urbanization traditionally exhibited a concentration of human activities and settlements around the downtown area, the so-called *in-migration*. In-migration refers to migration from former colonies and similar places. The fact that many immigrants settle in impoverished city centres led to the notion of the "peripheralization of the core", which simply describes that people who used to be at the periphery of the former empires now live right in the centre.

Recent developments, such as inner-city redevelopment schemes, mean that new arrivals in cities no longer necessarily settle in the centre. In some developed regions, the reverse effect, originally called counter urbanisation has occurred, with cities losing population to rural areas, and is particularly common for richer families. This has been possible because of improved communications, and has been caused by factors such as the fear of crime and poor urban environments. Later termed *"white flight"*, the effect is not restricted to cities with a high ethnic minority population.

When the residential area shifts outward, this is called suburbanization. A number of researchers and writers suggest that suburbanization has gone so far to form new points of concentration outside the downtown both in developed and developing countries such as India. This networked, poly-centric form of concentration is considered by some an emerging pattern of urbanization. It is called variously exurbia, edge city (Garreau, 1991), network city (Batten, 1995), or postmodern city (Dear, 2000). Los Angeles is the best-known example of this type of urbanization.

Rural migrants are attracted by the possibilities that cities can offer, but often settle in shanty towns and experience extreme poverty. In the 1980s, this was attempted to be tackled with the urban bias theory which was promoted by Michael Lipton who wrote: "...the most important class conflict in the poor countries of the world today is not between labour and capital. Nor is it between foreign and national interests. It is between rural classes and urban classes. The rural sector contains most of the poverty and most of the low-cost sources of potential advance; but the urban sector contains most of the articulateness, organization and power. So the urban classes have been able to win most of the rounds of the struggle with the countryside...". Most of the urban poor in developing countries able to find work can spend their lives in insecure, poorly paid jobs. According to research by the Overseas Development Institute pro-poor urbanisation will require labour intensive growth, supported by labour protection, flexible land use regulation and investments in basic services.'

Urbanization can be planned urbanization or organic. Planned urbanization, ie: planned community or the garden city movement, is based on an advance plan, which can be prepared for military, aesthetic, economic or urban design reasons. Examples can be seen in many ancient cities; although with exploration came the collision of nations, which meant that many invaded cities took on the desired planned characteristics of their occupiers. Many ancient organic cities experienced redevelopment for military and economic purposes, new roads carved through the cities, and new parcels of land were cordoned off serving various planned purposes giving cities distinctive geometric designs. UN agencies prefer to see urban infrastructure installed before urbanization occurs. Landscape planners are responsible for landscape infrastructure (public parks, sustainable urban drainage systems, greenways etc) which can be planned before urbanization takes place, or afterward to revitalize an area and create greater livability within a region. Concepts of control of the urban expansion are considered in the American Institute of Planners.

2

Geography of Europe

Europe is, by convention, one of the world's seven continents. Comprising the westernmost peninsula of Eurasia, Europe is generally divided from Asia to its east by the water divide of the Ural Mountains, the Ural River, the Caspian Sea, the Caucasus region (Specification of borders) and the Black Sea to the southeast. Europe is bordered by the Arctic Ocean and other bodies of water to the north, the Atlantic Ocean to the west, the Mediterranean Sea to the south, and the Black Sea and connected waterways to the southeast. Yet the borders for Europe—a concept dating back to classical antiquity—are somewhat arbitrary, as the term *continent* can refer to a cultural and political distinction or a physiographic one.

Europe is the world's second-smallest continent by surface area, covering about 10,180,000 square kilometres (3,930,000 sq mi) or 2% of the Earth's surface and about 6.8% of its land area. Of Europe's approximately 50 states, Russia is the largest by both area and population (although the country covers both Europe and Asia), while the Vatican City is the smallest. Europe is the third-most populous continent after Asia and Africa, with a population of 731 million or about 11% of the world's population. However, according to the United Nations (medium estimate), Europe's share may fall to about 7% by 2050. In 1900, Europe's share of the world's population was 25%.

Europe, in particular Ancient Greece, is the birthplace of Western culture. It played a predominant role in global affairs from the 16th century onwards, especially after the beginning of colonialism. Between the 16th and 20th centuries, European nations controlled at various times the Americas, most of Africa, Oceania, and large portions of Asia. Both World Wars were largely focused upon Europe, greatly contributing to a decline in Western European dominance in world affairs by the mid-20th century as the United States and Soviet Union took prominence. During the Cold War, Europe was divided along the Iron Curtain between NATO in the west and the Warsaw Pact in the east. European integration led to the formation of the Council of Europe and the European Union in Western Europe, both of which have been expanding eastward since the fall of the Soviet Union in 1991.

Definition

The use of the term "Europe" has developed gradually throughout history. In antiquity, the Greek historian Herodotus mentioned that the world had been divided by unknown persons into the three continents of Europe, Asia, and Libya (Africa), with the Nile and the river Phasis forming their boundaries—though he also states that some considered the River Don, rather than the Phasis, as the boundary between Europe and Asia. Flavius Josephus and the *Book of Jubilees* described the continents as the lands given by Noah to his three sons; Europe was defined as between the Pillars of Hercules at the Strait of Gibraltar, separating it from Africa, and the Don, separating it from Asia.

A cultural definition of Europe as its being the lands of Latin Christendom coalesced in the eighth century, signifying the new cultural condominium created through the confluence of Germanic traditions and Christian-Latin culture, defined partly in contrast with Byzantium and Islam, and limited to northern Iberia, the British Isles, France, Christianized western Germany, the Alpine regions and northern and central Italy. This division—as much cultural as geographical—was used until the Late Middle Ages, when it was challenged by the Age of Discovery. The problem of redefining Europe was finally resolved in 1730 when, instead of waterways, the Swedish geographer and cartographer von Strahlenberg proposed the Ural Mountains as the most significant eastern boundary, a suggestion that found favour in Russia and throughout Europe. Europe is now generally defined by geographers as the westernmost peninsula of Eurasia, with its boundaries marked by large bodies of water to the north, west and south; Europe's limits to the far east are usually taken to be the Urals, the Ural River, and the Caspian Sea; to the south-east, the Caucasus Mountains, the Black Sea and the waterways connecting the Black Sea to the Mediterranean Sea.

Sometimes, the word 'Europe' is used in a geopolitically limiting way to refer only to the European Union or, even more exclusively, a culturally defined core. On the other hand, the Council of Europe has 47 member countries, and only 27 member states are in the EU. In addition, people living in insular areas such as Ireland, the United Kingdom, the North Atlantic and Mediterranean islands and also in Scandinavia may routinely refer to "continental" or "mainland" Europe simply as Europe or "the Continent".

Etymology

In ancient Greek mythology, Europa was a Phonation princess whom Zeus abducted after assuming the form of a dazzling white bull. He took her to the island of Crete where she gave birth to Minos, Rhadamanthus and Sarpedon. For Homer, Europe was a mythological queen of Crete, not a geographical designation. Later, *Europa* stood for central-north Greece, and by 500 BC its meaning had been extended to the lands to the north.

The name of *Europa* is of uncertain etymology. One theory suggests that it is derived from the Greek roots meaning broad (*eur-*) and eye, hence *Europe*, "wide-gazing", "broad of aspect" (compare with *glaukopis* (grey-eyed) Athena or *boopis* (ox-eyed) Hera). *Broad* has been an epithet of Earth itself in the reconstructed Proto-Indo-European religion. Another theory suggests that it is actually based on a Semitic word such as the Akkadian *erebu* meaning "to go down, set" cognate to Phonation *'ereb* "evening; west" and Arabic Maghreb, Hebrew *ma'ariv*. However, M. L. West states that "phonologically, the match between Europa's name and any form of the Semitic word is very poor".

Most major world languages use words derived from "Europa" to refer to the "continent" (peninsula). Chinese, for example, uses the word *Ouzhou* (Pk2m), which is an abbreviation of the transliterated name *Ouluoba zhou*; however, in some

Turkic languages the name *Frengistan* (land of the Franks) is used casually in referring to much of Europe, besides official names such as *Avrupa* or *Evropa*.

History

Prehistory

Homo georgicus, which lived roughly 1.8 million years ago in Georgia, is the earliest hominid to have been discovered in Europe. Other hominid remains, dating back roughly 1 million years, have been discovered in Atapuerca, Spain. Neanderthal man (named for the Neander Valley in Germany) appeared in Europe 150,000 years ago and disappeared from the fossil record about 30,000 years ago. The Neanderthals were supplanted by modern humans (Cro-Magnons), who appeared in Europe around 40,000 years ago.

The European Neolithic period—marked by the cultivation of crops and the raising of livestock, increased numbers of settlements and the widespread use of pottery—began around 7,000 BC in Greece and the Balkans, probably influenced by earlier farming practices in Anatolia and the Near East. It spread from South Eastern Europe along the valleys of the Danube and the Rhine (Linear Pottery culture) and along the Mediterranean coast (Cardial culture). Between 4,500 and 3,000 BC these central European neolithic cultures developed further to the west and the north, transmitting newly acquired skills in producing copper artefacts. In Western Europe the Neolithic period was characterised not by large agricultural settlements but by field monuments, such as cause wayed enclosures, burial mounds and megalithic tombs. The Corded ware cultural horizon flourished at the transition from the Neolithic to the Chalcolithic. During this period giant megalithic monuments, such as the Megalithic Temples of Malta and Stonehenge, were constructed throughout Western and Southern Europe. The European Bronze Age began in the late 3rd millennium BC with the Beaker culture.

The European Iron Age began around 800 BC, with the Hallstatt culture. Iron Age colonisation by the Phoenicians gave rise to early Mediterranean cities. Early Iron Age Italy and Greece from around the 8th century BC gradually gave rise to historical Classical Antiquity.

Classical Antiquity

Ancient Greece had a profound impact on Western civilisation. Western democratic and individualistic culture are often attributed to Ancient Greece. The Greeks invented the polis, or city-state, which played a fundamental role in their concept of identity. These Greek political ideals were rediscovered in the late 18th century by European philosophers and idealists. Greece also generated many cultural contributions: in philosophy, humanism and rationalism under Aristotle, Socrates and Plato; in history with Herodotus and Thucydides; in dramatic and narrative verse, starting with the epic poems of Homer; and in science with Pythagoras, Euclid and Archimedes.

Another major influence on Europe came from the Roman Empire which left its mark on law, language, engineering, architecture, and government. During the *pax romana*, the Roman Empire expanded to encompass the entire Mediterranean Basin and much of Europe.

Stoicism influenced Roman emperors such as Hadrian, Antoninus Pius, and Marcus Aurelius, who all spent time on the Empire's northern border hting Germanic, Pictish and Scottish tribes. Christianity was eventually legitimised by Constantine I after three centuries of imperial persecution.

Early Middle Ages

During the decline of the Roman Empire, Europe entered a long period of change arising from what historians call the "Age of Migrations". There were numerous invasions and migrations amongst the Ostrogoths, Visigoths, Goths, Vandals, Huns, Franks, Angles, Saxons, Slavs, Avars, Bulgars and, later still, the Vikings and Magyars. Renaissance thinkers such as Petrarch would later refer to this as the "Dark Ages". Isolated monastic communities were the only places to safeguard and compile written knowledge accumulated previously; apart from this very few written records survive and much literature, philosophy, mathematics, and other thinking from the classical period disappeared from Europe.

During the Dark Ages, the Western Roman Empire fell under the control of various barbarian tribes. The Germanic and Slav tribes established their domains over Western and

Eastern Europe respectively. Eventually the Frankish tribes were united under Clovis I. Charlemagne, a Frankish king of the Carolingian dynasty who had conquered most of Western Europe, was anointed "Holy Roman Emperor" by the Pope in 800. This led to the founding of the Holy Roman Empire, which eventually became centred in the German principalities of central Europe.

The predominantly Greek speaking Eastern Roman Empire became known in the west as the Byzantine Empire. Its capital was Constantinople. Emperor Justinian I presided over Constantinople's first golden age: he established a legal code, funded the construction of the Hagia Sophia and brought the Christian church under state control. Fatally weakened by the sack of Constantinople during the Fourth Crusade, the Byzantines fell in 1453 when they were conquered by the Ottoman Empire.

Middle Ages

The Middle Ages were dominated by the two upper echelons of the social structure: the nobility and the clergy. Feudalism developed in France in the Early Middle Ages and soon spread throughout Europe. A struggle for influence between the nobility and the monarchy in England led to the writing of the Magna Carta and the establishment of a parliament. The primary source of culture in this period came from the Roman Catholic Church. Through monasteries and cathedral schools, the Church was responsible for education in much of Europe.

The Papacy reached the height of its power during the High Middle Ages. A East-West Schism in 1054 split the former Roman Empire religiously, with the Eastern Orthodox Church in the Byzantine Empire and the Roman Catholic Church in the former Western Roman Empire. In 1095 Pope Urban II called for a crusade against Muslims occupying Jerusalem and the Holy Land. In Europe itself, the Church organised the Inquisition against heretics. In Spain, the Reconquista concluded with the fall of Granada in 1492, ending over seven centuries of Muslim rule in the Iberian Peninsula.

In the 11th and 12th centuries, constant incursions by nomadic Turkic tribes, such as the Pechenegs and the Kipchaks, caused a massive migration of Slavic populations to the safer, heavily forested regions of the north. Like many other parts of Eurasia, these territories were overrun by the Mongols. The invaders, later known as Tatars, formed the state of the Golden Horde, which ruled the southern and central expanses of Russia for over three centuries.

The Great Famine of 1315–1317 was the first crisis that would strike Europe in the late Middle Ages. The period between 1348 and 1420 witnessed the heaviest loss. The population of France was reduced by half. Medieval Britain was afflicted by 95 famines, and France suffered the effects of 75 or more in the same period. Europe was devastated in the mid-14th century by the Black Death, one of the most deadly pandemics in human history which killed an estimated 25 million people in Europe alone—a third of the European population at the time.

The plague had a devastating effect on Europe's social structure; it induced people to live for the moment as illustrated by Giovanni Boccaccio in *The Decameron* (1353). It was a serious blow to the Roman Catholic Church and led to increased persecution of Jews, foreigners, beggars and lepers. The plague is thought to have returned every generation with varying virulence and mortalities until the 1700s. During this period, more than 100 plague epidemics swept across Europe.

Early Modern Period

The Renaissance was a period of cultural change originating in Florence and later spreading to the rest of Europe. in the fourteenth century. The rise of a new humanism was accompanied by the recovery of forgotten classical and Arabic knowledge from monastic libraries and the Islamic world. The Renaissance spread across Europe between the 14th and 16th centuries: it saw the flowering of art, philosophy, music, and the sciences, under the joint patronage of royalty, the nobility, the Roman Catholic Church, and an emerging merchant class. Patrons in Italy, including the Medici family of Florentine bankers and the Popes in Rome, funded prolific quattrocento and cinquecento artists such as Raphael, Michelangelo, and Leonardo da Vinci.

Political intrigue within the Church in the mid-14th century caused the Great Schism. During this forty-year period, two popes—one in Avignon and one in Rome—claimed rulership over the Church. Although the schism was eventually healed in 1417, the papacy's spiritual authority had suffered greatly.

The Church's power was further weakened by the Protestant Reformation (1517–1648), initially sparked by the works of German theologian Martin Luther, a result of the lack of reform within the Church. The Reformation also damaged the Holy Roman Empire's power, as German princes became divided between Protestant and Roman Catholic faiths. This eventually led to the Thirty Years War (1618–1648), which crippled the Holy Roman Empire and devastated much of Germany, killing between 25 and 40 percent of its population. In the aftermath of the Peace of Westphalia, France rose to predominance within Europe. The 17th century in southern and eastern Europe was a period of general decline. Eastern Europe experienced more than 150 famines in a 200-year period between 1501 to 1700.

The Renaissance and the New Monarchs marked the start of an Age of Discovery, a period of exploration, invention, and scientific development. In the 15th century, Portugal and Spain, two of the greatest naval powers of the time, took the lead in exploring the world. Christopher Columbus reached the New World in 1492, and soon after the Spanish and Portuguese began establishing colonial empires in the Americas. France, the Netherlands and England soon followed in building large colonial empires with vast holdings in Africa, the Americas, and Asia.

18th and 19th Centuries

The Age of Enlightenment was a powerful intellectual movement during the eighteenth century promoting scientific and reason-based thoughts. Discontent with the aristocracy and clergy's monopoly on political power in France resulted in the French Revolution and the establishment of the First Republic as a result of which the monarchy and many of the nobility perished during the initial reign of terror. Napoleon Bonaparte rose to power in the aftermath of the French Revolution and established the First French Empire that, during the Napoleonic Wars, grew to encompass large parts of Europe before collapsing in 1815 with the Battle of Waterloo.

Napoleonic rule resulted in the further dissemination of the ideals of the French Revolution, including that of the nation-state, as well as the widespread adoption of the French models of administration, law, and education. The Congress of Vienna, convened after Napoleon's downfall, established a new balance of power in Europe centred on the five "Great Powers": the United Kingdom, France, Prussia, Habsburg Austria, and Russia.

This balance would remain in place until the Revolutions of 1848, during which liberal uprisings affected all of Europe except for Russia and Great Britain. These revolutions were eventually put down by conservative elements and few reforms resulted. In 1867, the Austro-Hungarian empire was formed; and 1871 saw the unifications of both Italy and Germany as nation-states from smaller principalities.

The Industrial Revolution started in Great Britain in the last part of the 18th century and spread throughout Europe. The invention and implementation of new technologies resulted in rapid urban growth, mass employment, and the rise of a new working class. Reforms in social and economic spheres followed, including the first laws on child labour, the legalisation of trade unions, and the abolition of slavery. In Britain, the Public Health Act 1875 was passed, which significantly improved living conditions in many British cities. Europe's population doubled during the 18th century, from roughly 100 million to almost 200 million, and doubled again during the 19th century. In the 19th century, 70 million people left Europe in migrations to various European colonies abroad and to the United States.

20th Century to Present

Two World Wars and an economic depression dominated the first half of the 20th century. World War I was fought between 1914 and 1918. It started when Archduke Franz Ferdinand of Austria was assassinated by the Bosnian Serb Gavrilo Princip. Most European nations were drawn into the war, which was fought between the Entente Powers (France, Belgium, Serbia, Portugal, Russia, the United Kingdom, and later Italy, Greece, Romania, and the United States) and the Central Powers (Austria-Hungary, Germany, Bulgaria, and the Ottoman Empire). The War left around 40 million civilians and military dead. Over 60 million European soldiers were mobilised from 1914–1918.

Partly as a result of its defeat Russia was plunged into the Russian Revolution, which threw down the Tsarist monarchy and replaced it with the communist Soviet Union. Austria-Hungary and the Ottoman Empire collapsed and broke up into separate nations, and many other nations had their borders redrawn. The Treaty of Versailles, which officially ended World War I in 1919, was harsh towards Germany, upon whom it placed full responsibility for the war and imposed heavy sanctions.

Economic instability, caused in part by debts incurred in the First World War and 'loans' to Germany played havoc in Europe in the late 1920s and 1930s. This and the Wall Street Crash of 1929 brought about the worldwide Great Depression. Helped by the economic crisis, social instability and the threat of communism, fascist movements developed throughout Europe placing Adolf Hitler of Nazi Germany, Francisco Franco of Spain and Benito Mussolini of Italy in power.

In 1933, Hitler became the leader of Germany and began to work towards his goal of building Greater Germany. Germany re-expanded and took back the Saarland and Rhineland in 1935 and 1936. In 1938, Austria became a part of Germany too, following the Anschluss. Later that year, Germany annexed the German Sudetenland, which had become a part of Czechoslovakia after the war. This move was highly contested by the other powers, but ultimately permitted in the hopes of avoiding war and appeasing Hitler.

Shortly afterwards, Poland and Hungary started to press for the annexation of parts of Czechoslovakia with Polish and Hungarian majorities. Hitler encouraged the Slovaks to do the same and in early 1939, the remainder of Czechoslovakia was split into the Protectorate of Bohemia and Moravia, controlled by Germany, and the Slovak Republic, while other smaller regions went to Poland and Hungary. With tensions mounting between Germany and Poland over the future of Danzig, the Germans turned to the Soviets, and signed the Molotov-Ribbentrop Pact. Germany invaded Poland on 1 September 1939, prompting France and the United Kingdom to declare war on Germany on 3 September, opening the European theatre of World War II. The Soviet invasion of Poland started on 17 September and Poland fell soon thereafter.

On 24 September, the Soviet Union attacked the Baltic countries and later, Finland. The British hoped to land at Narvik and send troops to aid Finland, but their primary objective in the landing was to encircle Germany and cut the Germans off from Scandinavian resources. Nevertheless, the Germans knew of Britain's plans and got to Narvik first, repulsing the attack. Around the same time, Germany moved troops into Denmark, which left no room for a front except for where the last war had been fought or by landing at sea. The Phoney War continued.

In May 1940, Germany attacked France through the Low Countries. France capitulated in June 1940. However, the British refused to negotiate peace terms with the Germans and the war continued. By August Germany began a bombing offensive on Britain, but failed to convince the Britons to give up. In 1941, Germany invaded the Soviet Union in the ultimately unsuccessful Operation Barbarossa. On 7 December 1941 Japan's attack on Pearl Harbor drew the United States into the conflict as allies of the British Empire and other allied forces.

After the staggering Battle of Stalingrad in 1943, the German offensive in the Soviet Union turned into a continual fallback. In 1944, British and American forces invaded France in the D-Day landings, opening a new front against Germany. Berlin finally fell in 1945, ending World War II in Europe.

The war was the largest and most destructive in human history, with 60 million dead across the world. More than 40 million people in Europe had lost their lives by the time World War II ended, including between 11 and 17 million people who perished during the Holocaust. The Soviet Union lost around 27 million people during the war, about half of all World War II casualties. By the end of World War II, Europe had more than 40 million refugees. Several post-war expulsions in Central and Eastern Europe displaced a total of about 20 million people.

World War I and especially World War II diminished the eminence of Western Europe in world affairs. After World War II the map of Europe was redrawn at the Yalta Conference and divided into two blocs, the Western countries and the communist Eastern bloc, separated by what was later called by Winston

Churchill an "iron curtain". The United States and Western Europe established the NATO alliance and later the Soviet Union and Eastern Europe established the Warsaw Pact.

The two new superpowers, the United States and the Soviet Union, became locked in a fifty-year long Cold War, centred on nuclear proliferation. At the same time decolonisation, which had already started after World War I, gradually resulted in the independence of most of the European colonies in Asia and Africa. In the 1980s the reforms of Mikhail Gorbachev and the Solidarity movement in Poland accelerated the collapse of the Eastern bloc and the end of the Cold War. Germany was reunited, after the symbolic fall of the Berlin Wall in 1989, and the maps of Eastern Europe were redrawn once more. European integration also grew in the post-World War II years. The Treaty of Rome in 1957 established the European Economic Community between six Western European states with the goal of a unified economic policy and common market. In 1967 the EEC, European Coal and Steel Community and Euratom formed the European Community, which in 1993 became the European Union. The EU established a parliament, court and central bank and introduced the euro as a unified currency. In 2004 and 2007, Eastern European countries began joining, expanding the EU to its current size of 27 European countries, and once more making Europe a major economical and political centre of power.

Geography and Extent

Physiographically, Europe is the northwestern constituent of the larger landmass known as Eurasia, or Afro-Eurasia: Asia occupies the eastern bulk of this continuous landmass and all share a common continental shelf. Europe's eastern frontier is now commonly delineated by the Ural Mountains in Russia.

The first border definition was intoduced in 5th Century B.C. by the "father of history" Herodotus, when he regarded Europe to be extending to the Eastern Ocean, and being as long as (and much larger than) Africa and Asia together. He marked the borders between Europe and Asia on Kura River and Rioni River in Transcaucasia. The first century AD geographer Strabo, took the River Don "*Tanais*" to be the boundary to the Black Sea, as did early Judaic sources.

The southeast boundary with Asia is not universally defined, with the Ural River, or alternatively, the Emba River most commonly serving as possible boundaries. The boundary continues to the Caspian Sea, the crest of the Caucasus Mountains or, alternatively, the Kura River in the Caucasus, and on to the Black Sea; the Bosporus, the Sea of Marmara, the Dardanelles, and the Aegean Sea conclude the Asian boundary. The Mediterranean Sea to the south separates Europe from Africa. The western boundary is the Atlantic Ocean; Iceland, though nearer to Greenland (North America) than mainland Europe, is generally included in Europe.

Because of sociopolitical and cultural differences, there are various descriptions of Europe's boundary; in some sources, some territories are not included in Europe, while other sources include them. For instance, geographers from Russia and other post-Soviet states generally include the Urals in Europe while including Caucasia in Asia. Similarly, Cyprus is approximate to Anatolia (or Asia Minor), but is often considered part of Europe and currently is a member state of the EU. In addition, Malta was considered an island of Africa for centuries.

Physical Geography

Land relief in Europe shows great variation within relatively small areas. The southern regions are more mountainous, while moving north the terrain descends from the high Alps, Pyrenees and Carpathians, through hilly uplands, into broad, low northern plains, which are vast in the east. This extended lowland is known as the Great European Plain, and at its heart lies the North German Plain. An arc of uplands also exists along the north-western seaboard, which begins in the western parts of the islands of Britain and Ireland, and then continues along the mountainous, fjord-cut, spine of Norway.

This description is simplified. Sub-regions such as the Iberian Peninsula and the Italian Peninsula contain their own complex features, as does mainland Central Europe itself, where the relief contains many plateaus, river valleys and basins that complicate the general trend. Sub-regions like Iceland, Britain and Ireland are special cases. The former is a land unto itself in the northern ocean which is counted as part of Europe, while the latter are upland areas that were once joined to the mainland until rising sea levels cut them off.

Climate

The climate is milder in comparison to other areas of the same latitude around the globe due to the influence of the Gulf Stream. The Gulf Stream is nicknamed "Europe's central heating", because it makes Europe's climate warmer and wetter than it would otherwise be. The Gulf Stream not only carries warm water to Europe's coast but also warms up the prevailing westerly winds that blow across the continent from the Atlantic Ocean.

Therefore the average temperature throughout the year of Naples is 16 °C (60.8 °F), while it is only 12 °C (53.6 °F) in New York City which is almost on the same latitude. Berlin, Germany; Calgary, Canada; and Irkutsk, in the Asian part of Russia, lie on around the same latitude; January temperatures in Berlin average around 8 °C (15 °F) higher than those in Calgary, and they are almost 22 °C (40 °F) higher than average temperatures in Irkutsk.

Geology

The Geology of Europe is hugely varied and complex, and gives rise to the wide variety of landscapes found across the continent, from the Scottish Highlands to the rolling plains of Hungary.

Europe's most significant feature is the dichotomy between highland and mountainous Southern Europe and a vast, partially underwater, northern plain ranging from Ireland in the west to the Ural Mountains in the east. These two halves are separated by the mountain chains of the Pyrenees and Alps/Carpathians. The northern plains are delimited in the west by the Scandinavian Mountains and the mountainous parts of the British Isles. Major shallow water bodies submerging parts of the northern plains are the Celtic Sea, the North Sea, the Baltic Sea complex and Barents Sea.

The northern plain contains the old geological continent of Baltica, and so may be regarded geologically as the "main continent", while peripheral highlands and mountainous regions in the south and west constitute fragments from various other geological continents. Most of the older geology of Western Europe existed as part of the ancient microcontinent Avalonia.

Geological History

The geological history of Europe traces back to the formation of the Baltic Shield (Fennoscandia) and the Sarmatian craton, both around 2.25 billion years ago, followed by the Volgo-Uralia shield, the three together leading to the East European craton (H" Baltica) which became a part of the supercontinent Columbia. Around 1.1 billion years ago, Baltica and Arctica (as part of the Laurentia block) became joined to Rodinia, later resplitting around 550 million years ago to reform as Baltica. Around 440 million years ago Euramerica was formed from Baltica and Laurentia; a further joining with Gondwana then leading to the formation of Pangea. Around 190 million years ago, Gondwana and Laurasia split apart due to the widening of the Atlantic Ocean. Finally, and very soon afterwards, Laurasia itself split up again, into Laurentia (North America) and the Eurasian continent. The land connection between the two persisted for a considerable time, via Greenland, leading to interchange of animal species. From around 50 million years ago, rising and falling sea levels have determined the actual shape of Europe, and its connections with continents such as Asia. Europe's present shape dates to the late Tertiary period about five million years ago.

Biodiversity

Having lived side-by-side with agricultural peoples for millennia, Europe's animals and plants have been profoundly affected by the presence and activities of man. With the exception of Fennoscandia and northern Russia, few areas of untouched wilderness are currently found in Europe, except for various national parks.

The main natural vegetation cover in Europe is mixed forest. The conditions for growth are very favourable. In the north, the Gulf Stream and North Atlantic Drift warm the continent. Southern Europe could be described as having a warm, but mild climate. There are frequent summer droughts in this region. Mountain ridges also affect the conditions. Some of these (Alps, Pyrenees) are oriented east-west and allow the wind to carry large masses of water from the ocean in the interior. Others are oriented south-north (Scandinavian Mountains, Dinarides, Carpathians, Apennines) and because the rain falls primarily on the side of mountains that is oriented towards sea, forests grow well on this side, while on the other side, the conditions are much less favourable. Few corners of mainland Europe have not been grazed by livestock at some point in time, and the cutting down of the pre-agricultural forest habitat caused disruption to the original plant and animal ecosystems.

Probably eighty to ninety per cent of Europe was once covered by forest. It stretched from the Mediterranean Sea to the Arctic Ocean. Though over half of Europe's original forests disappeared through the centuries of deforestation, Europe still has over one quarter of its land area as forest, such as the taiga of Scandinavia and Russia, mixed rainforests of the Caucasus and the Cork oak forests in the western Mediterranean. During recent times, deforestation has been slowed and many trees have been planted. However, in many cases monoculture plantations of conifers have replaced the original mixed natural forest, because these grow quicker. The plantations now cover vast areas of land, but offer poorer habitats for many European forest dwelling species which require a mixture of tree species and diverse forest structure. The amount of natural forest in Western Europe is just 2–3% or less, in European Russia 5–10%. The country with the smallest percentage of forested area (excluding the micronations) is Iceland (1%), while the most forested country is Finland (77%).

In temperate Europe, mixed forest with both broadleaf and coniferous trees dominate. The most important species in central and western Europe are beech and oak. In the north, the taiga is a mixed spruce–pine–birch forest; further north within Russia and extreme northern Scandinavia, the taiga gives way to tundra as the Arctic is approached. In the Mediterranean, many olive trees have been planted, which are very well adapted to its arid climate; Mediterranean Cypress is also widely planted in southern Europe. The semi-arid Mediterranean region hosts much scrub forest. A narrow east-west tongue of Eurasian grassland (the steppe) extends eastwards from Ukraine and southern Russia and ends in Hungary and traverses into taiga to the north. Glaciation during the most recent ice age and the presence of man affected the distribution of European fauna.

As for the animals, in many parts of Europe most large animals and top predator species have been hunted to extinction. The woolly mammoth was extinct before the end of the Neolithic period. Today wolves (carnivores) and bears (omnivores) are endangered. Once they were found in most parts of Europe. However, deforestation and hunting caused these animals to withdraw further and further. By the Middle Ages the bears' habitats were limited to more or less inaccessible mountains with sufficient forest cover. Today, the brown bear lives primarily in the Balkan peninsula, Scandinavia, and Russia; a small number also persist in other countries across Europe (Austria, Pyrenees etc.), but in these areas brown bear populations are fragmented and marginalised because of the destruction of their habitat. In addition, polar bears may be found on Svalbard, a Norwegian archipelago far north of Scandinavia. The wolf, the second largest predator in Europe after the brown bear, can be found primarily in Eastern Europe and in the Balkans, with a handful of packs in pockets of Western Europe (Scandinavia, Spain, etc.).

European wild cat, foxes (especially the red fox), jackal and different species of martens, hedgehogs, different species of reptiles (like snakes such as vipers and grass snakes) and amphibians, different birds (owls, hawks and other birds of prey).

Important European herbivores are snails, larvae, fish, different birds, and mammals, like rodents, deer and roe deer, boars, and living in the mountains, marmots, steinbocks, chamois among others.

The extinction of the dwarf hippos and dwarf elephants has been linked to the earliest arrival of humans on the islands of the Mediterranean.

Sea creatures are also an important part of European flora and fauna. The sea flora is mainly phytoplankton. Important animals that live in European seas are zooplankton, molluscs, echinoderms, different crustaceans, squids and octopuses, fish, dolphins, and whales.

Biodiversity is protected in Europe through the Council of Europe's Bern Convention, which has also been signed by the European Community as well as non-European states.

Demographics

Since the Renaissance, Europe has had a major influence in culture, economics and social movements in the world. The most significant inventions had their origins in the Western world, primarily Europe and the United States. Some current and past issues in European demographics have included religious emigration, race relations, economic immigration, a declining birth rate and an aging population.

In some countries, such as Ireland and Poland, access to abortion is currently limited; in the past, such restrictions and also restrictions on artificial birth control were commonplace throughout Europe. Abortion remains illegal on the island of Malta where Catholicism is the state religion. Furthermore, three European countries (the Netherlands, Belgium, and Switzerland) and the Autonomous Community of Andalusia (Spain) have allowed a limited form of voluntary euthanasia for some terminally ill people.

In 2005, the population of Europe was estimated to be 731 million according to the United Nations, which is slightly more than one-ninth of the world's population. A century ago, Europe had nearly a quarter of the world's population. The population of Europe has grown in the past century, but in other areas of the world (in particular Africa and Asia) the population has grown far more quickly. Among the continents, Europe has a relatively high population density, second only to Asia. The most densely populated country in Europe is the Netherlands, ranking third in the world after Bangladesh and South Korea. Pan and Pfeil (2004) count 87 distinct "peoples of Europe", of which 33 form the majority population in at least one sovereign state, while the remaining 54 constitute ethnic minorities.

According to UN population projection, Europe's population may fall to about 7% of world population by 2050, or 653 million people (medium variant, 556 to 777 million in low and high variants, respectively). Within this context, significant disparities exist between regions in relation to fertility rates. The average number of children per female of child bearing age is 1.52. According to some sources, this rate is higher among Muslims in Europe. The UN predicts the steady population decline of vast areas of Eastern Europe. The Russia's population is declining by at least 700,000 people each year. The country now has 13,000 uninhabited villages.

Europe is home to the highest number of migrants of all global regions at 70.6 million people, the IOM's report said. In 2005, the EU had an overall net gain from immigration of 1.8 million people, despite having one of the highest population densities in the world. This accounted for almost 85% of Europe's total population growth. The European Union plans to open the job centres for legal migrant workers from Africa.

Emigration from Europe began with Spanish settlers in the 16th century, and French and English settlers in the 17th century. But numbers remained relatively small until waves of mass emigration in the 19th century, when millions of poor families left Europe.

Today, large populations of European descent are found on every continent. European ancestry predominates in North America, and to a lesser degree in South America (particularly in Argentina, Chile, Uruguay and Centro-Sul of Brazil). Also, Australia and New Zealand have large European derived populations. Africa has no countries with European-derived majorities, but there are significant minorities, such as the White South Africans. In Asia, European-derived populations (specifically Russians) predominate in Northern Asia.

Political Geography

According to different definitions, the territories may be subject to various categorisations. The 27 European Union member states are highly integrated economically and politically; the European Union itself forms part of the political geography of Europe. The table below shows the scheme for geographic subregions used by the United Nations, alongside the regional grouping published in the CIA factbook. The socio-geographical data included are per sources in cross-referenced articles.

Economy

As a continent, the economy of Europe is currently the largest on Earth and it is the richest region as measured by assets under management with over $32.7 trillion compared to North America's $27.1 trillion in 2008.In 2009 Europe remained the wealthiest region. Its $37.1 trillion in assets under management represented one-third of the world's wealth. It was one of several regions where wealth surpassed its precrisis year-end peak. As with other continents, Europe has a large variation of wealth among its countries. The richer states tend to be in the West; some of the Eastern economies are still emerging from the collapse of the Soviet Union and Yugoslavia.

The European Union, an intergovernmental body composed of 27 European states, comprises the largest single economic area in the world. Currently, 16 EU countries share the euro as a common currency. Five European countries rank in the top ten of the worlds largest national economies in GDP (PPP). This includes (ranks according to the CIA): Germany (5), the UK (6), Russia (7), France (8), and Italy (10).

Pre–1945: Industrial Growth

Capitalism has been dominant in the Western world since the end of feudalism. From Britain, it gradually spread throughout Europe. The Industrial Revolution started in Europe, specifically the United Kingdom in the late 18th century, and the 19th century saw Western Europe industrialise. Economies were disrupted by World War I but by the beginning of World War II they had recovered and were having to compete with the growing economic strength of the United States. World War II, again, damaged much of Europe's industries.

1945–1990: The Cold War

After World War II the economy of the UK was in a state of ruin, and continued to suffer relative economic decline in the following decades. Italy was also in a poor economic condition but regained a high level of growth by the 1950s. West Germany recovered quickly and had doubled production from prewar levels by the 1950s. France also staged a remarkable comeback enjoying rapid growth and modernisation; later on Spain, under the leadership of Franco, also recovered, and the nation recorded huge unprecedented economic growth beginning in the 1960s in what is called the Spanish miracle. The majority of Eastern European states came under the control of the USSR and thus were members of the Council for Mutual Economic Assistance (COMECON). The states which retained a free-market system were given a large amount of aid by the United States under the Marshall Plan. The western states moved to link their economies together, providing the basis for the EU and increasing cross border trade. This helped them to enjoy rapidly improving economies, while those states in COMECON were struggling in a large part due to the cost of the Cold War. Until 1990, the European Community was expanded from 6 founding members to 12. The emphasis placed on resurrecting the West German economy led to it overtaking the UK as Europe's largest economy.

1991–2007: Integration and Reunification

With the fall of communism in Eastern Europe in 1991 the Eastern states had to adapt to a free market system. There were varying degrees of success with Central European countries such as Poland, Hungary, and Slovenia adapting reasonably quickly, while eastern states like Ukraine and Russia taking far longer. Western Europe helped Eastern Europe by forming economic ties with it.

After East and West Germany were reunited in 1990, the economy of West Germany struggled as it had to support and largely rebuild the infrastructure of East Germany. Yugoslavia lagged farthest behind as it was ravaged by war and in 2003 there were still many EU and NATO peacekeeping troops in Kosovo, the Republic of Macedonia, and Bosnia and Herzegovina, with only Slovenia making any real progress.

By the millennium change, the EU dominated the economy of Europe comprising the five largest European economies of the time namely Germany, the United Kingdom, France, Italy, and Spain. In 1999 12 of the 15 members of the EU joined the Eurozone replacing their former national currencies by the common euro. The three who chose to remain outside the Eurozone were: the United Kingdom, Denmark, and Sweden.

2008–2010: Recession

The Eurozone entered its first official recession in the third quarter of 2008, official figures confirmed in January 2009. While beginning in the United States the late-2000s recession spread to Europe rapidly and has affected much of the region. The official unemployment rate in the 16 countries that use the euro rose to 9.5% in May 2009. Europe's young workers have been especially hard hit. In the first quarter of 2009, the unemployment rate in the EU27 for those aged 15–24 was 18.3%. In early 2010 fears of a sovereign debt crisis developed concerning some countries in Europe, especially Greece, Ireland, Spain, and Portugal. As a result, measures were taken especially for Greece by the leading countries of the Eurozone.

Language

European languages mostly fall within three Indo-European language groups: the Romance languages, derived from the Latin language of the Roman Empire; the Germanic languages, whose ancestor language came from southern Scandinavia; and the Slavic languages. While having the majority of its vocabulary descended from Romance languages, the English language is classified as a Germanic language.

Romance languages are spoken primarily in south-western Europe as well as in Romania and Moldova. Germanic languages are spoken in north-western Europe and some parts of Central Europe. Slavic languages are spoken in Central, Eastern, and Southeastern Europe.

Many other languages outside the three main groups exist in Europe. Other Indo-European languages include the Baltic group (i.e., Latvian and Lithuanian), the Celtic group (i.e., Irish, Scottish Gaelic, Manx, Welsh, Cornish, and Breton), Greek, Albanian, and Armenian. A distinct group of Uralic languages are Estonian, Finnish, and Hungarian, spoken in the respective countries as well as in parts of Romania, Russia, Serbia, and Slovakia. Other Non-Indo-European languages are Maltese (the only Semitic language official to the EU), Basque, Georgian, Azerbaijani, Turkish in Eastern Thrace, and the languages of minority nations in Russia.

Multilingualism and the protection of regional and minority languages are recognised political goals in Europe today. The Council of Europe Framework Convention for the Protection of National Minorities and the Council of Europe's European Charter for Regional or Minority Languages set up a legal framework for language rights in Europe.

Religion

Historically, religion in Europe has been a major influence on European art, culture, philosophy and law. The majority religion in Europe is Christianity as practiced by Catholic, Eastern Orthodox and Protestant Churches. Following these is Islam concentrated mainly in the south east (Bosnia and Herzegovina, Albania, Kosovo, Kazakhstan, North Cyprus, Turkey and Azerbaijan), and Tibetan Buddhism, found in Kalmykia. Other religions including Judaism and Hinduism are minority religions. Europe is a relatively secular continent and has the largest number and proportion of irreligious, agnostic and atheistic people in the Western world, with a particularly high number of self-described non-religious people in the Czech Republic, Estonia, Sweden, Germany (East), and France.

Culture

The culture of Europe can be described as a series of overlapping cultures; cultural mixes exist across the continent. There are cultural innovations and movements, sometimes at odds with each other. Thus the question of "common culture" or "common values" is complex.

Europe and its Geography

Europe is traditionally reckoned as one of seven continents. Physiographically, however, it is the northwestern peninsula of the larger landmass known as Eurasia (or Afro-Eurasia): Asia occupies the eastern bulk of this continuous landmass (save the Suez Canal separating Asia and Africa) and all share a common continental shelf. Europe's eastern frontier is delineated by the Ural Mountains in Russia. The south-east boundary with Asia is not universally defined. Most commonly the Ural or, alternatively, the Emba River serve as possible boundaries.

The boundary continues to the Caspian Sea, the crest of the Caucasus Mountains or, alternatively, the Kura River in the Caucasus, and on to the Black Sea; the Bosporus, the Sea of Marmara, and the Dardanelles conclude the Asian boundary. The Mediterranean Sea to the south separates Europe from Africa. The western boundary is the Atlantic Ocean; Iceland, though on the Mid-Atlantic Ridge and nearer to Greenland (North America) than mainland Europe, is generally included in Europe for cultural reasons. There is ongoing debate on where the geographical centre of Europe is.

Overview

The idea of a European "continent" is not universally held. Some geographical texts refer to a Eurasian Continent, or to a European subcontinent, given that Europe is not surrounded by sea and is, in any case, much more a cultural than a geographically definable area.

In terms of shape, Europe is a collection of connected peninsulas. The two largest of these are "mainland" Europe and Scandinavia to the north, divided from each other by the Baltic Sea. Three smaller peninsulas—Iberia, Italy and the Balkans—emerge from the southern margin of the mainland into the Mediterranean Sea, which separates Europe from Africa. Eastward, mainland Europe widens much like the mouth of a funnel, until the boundary with Asia is reached at the Ural Mountains.

Land relief in Europe shows great variation within relatively small areas. The southern regions are mountainous, while moving north the terrain descends from the high Alps, Pyrenees and Carpathians, through hilly uplands, into broad, low northern plains, which are vast in the east. An arc of uplands also exists along the northwestern seaboard, beginning in the western British Isles and continuing along the mountainous, fjord-cut spine of Norway.

This description is simplified. Sub-regions such as Iberia and Italy contain their own complex features, as does mainland Europe itself, where the relief contains many plateaus, river valleys and basins that complicate the general trend. Iceland and the British Isles are special cases. The former is a land unto itself in the northern ocean which is counted as part of Europe, while the latter are upland areas that were once joined to the mainland until rising sea levels cut them off.

The few generalizations that can be made about the relief of Europe make it less than surprising that the continent's many separate regions provided homes for many separate nations throughout history.

Geology

Europe's most significant feature is the dichotomy between highland and mountainous Southern Europe and a vast, partially underwater, northern plain ranging from England in the west to Ural Mountains in the east. These two halves are separated by the mountain chains of Pyrenees and Alps/ Carpathians.

The northern plains are delimited in the west by the Scandinavian Mountains and the mountainous parts of the British Isles. Major shallow water bodies submerging parts of the northern plains are the Celtic Sea, the North Sea, the Baltic Sea complex, and the Barents Sea.

The northern plain contains the old geological continent of Baltica, and so may be regarded as the "main continent", while peripheral highlands and mountainous regions in south and west constitute fragments from various other geological continents.

The geology of Europe is varied and complex, and gives rise to the wide variety of landscapes found across the continent, from the Scottish Highlands to the rolling plains of Hungary. Europe's most significant feature is the dichotomy between highland and mountainous Southern Europe and a vast, partially underwater, northern plain ranging from England in the west to the Ural Mountains in the east. These two halves are separated by the Pyrenees and the Alps-Carpathians mountain chain. The northern plains are delimited in the west by the Scandinavian Mountains and the mountainous parts of the British Isles. The southern mountainous region is bounded by the Mediterranean Sea. Major shallow water bodies submerging parts of the northern plains are the Celtic Sea the North Sea, the Baltic Sea complex and Barents Sea.

Components

Europe consists of the following cratons and terranes and microcontinents:

- The Baltica craton-Scandinavia, Balticum, Russia, and Northern Poland/Germany,
- Avalonian fragments-England, Ireland, Netherlands, Northern Germany, etc.,
- Laurentian (North American) fragments-Western Norway and Scotland,
- Gondwana Fragments-Spain, Italy??, Malta, possibly belonging to the Cimmerian Arc,

- Neo-Tethys Ocean floor-the Pyrenees/Alps/Apennine/ Balkan Alps/Carpathian complex,
- The Anatolian part of the Cimmerian Arc originating from Gondwana

Lakes and Inland Seas

International Lakes in Europe

- Lake Constance (Austria, Germany, Switzerland; Bodensee)
- Dojran Lake (Republic of Macedonia and Greece)
- Lake Geneva (France, Switzerland; Lac Léman or Lac de Geneve)
- Lake Lugano (Switzerland, Italy)
- Lake Maggiore (Switzerland, Italy; Lago Maggiore)
- Lake Neusiedl (Neusiedler See)/Ferto (Austria, Hungary)
- Lake Ohrid (Albania, Republic of Macedonia)
- Lake Peipsi-Pihkva (Estonia, Russia)
- Lake Great Prespa (Albania, Republic of Macedonia, Greece)
- Lake Small Prespa (Albania, Greece)
- Lake Skadar (Montenegro, Albania)
- Lake Vistytis (Lithuania, Russia)
- Lago di Lei (an artificial lake created by a dam; the waters are mostly in Italy but the dam is in Switzerland).

Major Islands

Iceland, Faroe Islands, Great Britain, Ireland, the rest of the British Isles, Balearic Islands, Corsica, Sardinia, Sicily, Malta, Ionian Islands, Crete, Aegean Islands, Aland Islands, Gotland, Saaremaa, Svalbard, Hinnoya, Senja, Zealand, Fyn and North Jutlandic Island.

Plains and Lowlands

- East European Plain, the largest landscape feature of Europe
- Northern European Lowlands
- Pannonian plain

- Meseta Central is a high plain (plateau) in central Spain (occupies roughly 40% of the country)
- Po Valley, also known as Padan Plain, between Alps and Apennines.

Mountain Ranges

Some of Europe's major mountain ranges are:

- Ural Mountains, used to separate Europe and Asia
- Caucasus Mountains, which also separate Europe and Asia, and is the namesake of the Caucasian race
- Carpathian Mountains, a major mountain range in Central and Southern Europe
- Alps, the famous mountains known for their spectacular slopes
- Apennines, which run through Italy
- Pyrenees, the natural border between France and Spain
- Cantabrian Mountains, which run across northern Spain
- Scandinavian Mountains, a mountain range which runs through the Scandinavian Peninsula, includes the Kjolen mountains
- Dinaric Alps, a mountain range in the Balkans
- Balkan mountains, a mountain range in central Balkans
- Scottish highlands (cairngorms, a 'low level' mountain range, in northern and central Scotland.
- pennines, very low level mountain range, subject to extreme glacial sculpting, in earlier ice ages, found in northern England.

Temperature and Precipitation

The high mountainous areas of Europe are colder and have higher precipitation than lower areas, as is true of mountainous areas in general. Europe has less precipitation in the east than in central and Western Europe. The temperature difference between summer and winter gradually increases from coastal northwest Europe to southeast inland Europe, ranging from Ireland, with a temperature difference of only 10 °C from the warmest to the coldest month, to the area north of the Caspian Sea, with a temperature difference of 40 °C. January average range from 13°C in Southern Greece to-20°C in northeastern part of European Russia. Western Europe and parts of Central Europe generally fall into the temperate maritime climate (Cfb), the southern part is mostly a Mediterranean climate (mostly Csa, smaller area with Csb), the north-central part and east into central Russia is mostly a humid continental climate (Dfb) and the northern part of the continent is a subarctic climate (Dfc). In the extreme northern part (northernmost Russia; Svalbard), bordering the Arctic Ocean, is tundra climate (Et). Mountain ranges, such as the Alps and the Carpathian mountains, have a highland climate with large variations according to altitude and latitude.

Landlocked Countries

- Andorra
- Armenia
- Austria
- Belarus
- Czech Republic
- Hungary
- Luxembourg
- Liechtenstein
- Republic of Macedonia (name is currently under dispute)
- Moldova
- San Marino
- Serbia
- Slovakia
- Switzerland
- Vatican.

Notes:

1. Liechtenstein is doubly landlocked.
2. Switzerland, Liechtenstein, Austria, Czech Republic, Slovakia, Kosovo, Hungary, Serbia, and Macedonia constitute a contiguous landlocked agglomeration of nine countries in Central Europe and the Balkans, stretching from Geneva all the way to Greece.
3. All other landlocked countries (Luxembourg, Andorra, Vatican, San Marino, Belarus, Moldova, Azerbaijan, and Kazakhstan) are "standalone" landlocked countries,

not bordering any other such *European* one (the emphasis is necessary, since Kazakhstan borders Turkmenistan, Uzbekistan, and Kyrgyzstan, thus forming a vast landlocked expanse in Central Asia).

Countries Consisting Solely of Islands or Parts of Islands

- Iceland
- United Kingdom
- Ireland
- Malta
- Isle of Man, Jersey and Guernsey (dependencies of British Crown)
- Faroe Islands (dependencies of Denmark).

Countries Bordering or Spanning Another Continent

Eurasia	Armenia, Azerbaijan, Georgia, Kazakhstan, Russia, Cyprus, Turkey, Greece (some Aegean islands)
Europe-Africa	Spain (Ceuta, Melilla and Canary Islands), Italy (Lampedusa and Lampione), Portugal (Madeira), France (Reunion)
Europe-South America	France (French Guiana)
Europe-North and Central America	Portugal (Western part of the Azores), France (Guadeloupe, Saint Barthélemy and Martinique), the Netherlands (Netherlands Antilles).

Countries that share a name with their capital;

- Luxembourg
- Monaco
- San Marino
- Vatican.

Countries whose capital is not their largest city;

Country	*Capital*	*Largest City*
Liechtenstein	Vaduz	Schaan
Malta	Valetta	Birkirkara
San Marino	San Marino	Serravalle
Switzerland	Bern	Zurich
Turkey	Ankara	Istanbul

Note: Italy's capital, Rome, is the country's largest city if only the municipality (*comune*) is considered. According to some definitions of urban zone and metropolitan areas, Milan's and Naples' metropolitan areas are larger than Rome's.

List of countries by the number of other countries they border;

14 Russia (Norway, Finland, Estonia, Latvia, Lithuania, Poland, Belarus, Ukraine, Georgia, Azerbaijan, Kazakhstan, China, Mongolia, and North Korea)

10 France

9 Germany

8 Austria, Serbia, Turkey

7 Hungary, Poland, Ukraine

6 Italy

5 Bulgaria, Croatia, Romania, Switzerland, Belarus, Spain, Kazakhstan, Slovakia, Macedonia

4 Belgium, Greece, Albania, Montenegro, Slovenia, The Czech Republic, Latvia, Lithuania, Georgia, Armenia, Azerbaijan

3 Finland, Norway, Luxembourg, Bosnia and Herzegovina

2 Sweden, Andorra, Liechtenstein, Estonia, Moldova, Netherlands

1 Denmark, Ireland, United Kingdom, Monaco, Portugal, Vatican City, San Marino, Gibraltar

0 Isle of Man, Jersey, Guernsey (dependencies of the British Crown), Faroe Islands (dependency of Denmark), Iceland, Malta.

Note: includes borders of overseas departments, for example: France's overseas departments and collectivities also share land borders with Brazil and Suriname (bordering French Guiana), and the Netherlands (bordering Saint-Martin).

3

The British Isles

The British Isles are a group of islands off the northwest coast of continental Europe that include the islands of Great Britain and Ireland, and over six thousand smaller islands. There are two sovereign states located on the islands: the United Kingdom of Great Britain and Northern Ireland (commonly known as the United Kingdom) and Ireland (also described as the Republic of Ireland). The British Isles also include three dependencies of the United Kingdom: the Isle of Man and, by tradition, the Bailiwick of Jersey and the Bailiwick of Guernsey in the Channel Islands, although the latter are not physically a part of the island group.

The oldest rocks in the group are in the north west of Scotland and Ireland and are 2,700 million years old. During the Silurian period the north-western regions collided with the south-east, which had been part of a separate continental landmass. The topography of the islands is modest in scale by global standards. Ben Nevis rises to a elevation of only 1,344 m (4,409 ft) and Lough Neagh, which is notably larger than other lakes on the isles, covers only 381 km^2 (147 m^2). The climate is temperate marine, with mild winters and warm wet summers. The North Atlantic Drift brings significant moisture and raises temperatures 11 °C (20 °F) above the global average for the latitude. This led to a landscape which was long dominated by temperate rainforest, although human activity has since cleared the vast majority of forest cover. The region was re-inhabited after the last glacial period of Quaternary glaciation, by 12,000 BC in Great Britain and 8000 BC in Ireland. At that time, Great Britain was a peninsula of the

European continent from which Ireland had become separated to form an island.

Celts were inhabiting the islands by the time of the Roman Empire, which expanded to control most of Britain. The first Anglo-Saxons arrived as Roman power waned in the 5th century, eventually occupying the bulk of what is now England. Viking invasions began in the ninth century, followed by more permanent settlements and political change-particularly in England. The subsequent Norman conquest of England in 1066 and the later Angevin conquest of Ireland from 1169 led to the imposition of a new Norman ruling elite across all of Britain and most of Ireland. By the Late Middle Ages, Great Britain was separated into the Kingdoms of England and Scotland, while control in Ireland fluxed between Gaelic kingdoms, Hiberno-Norman lords and the English-dominated Lordship of Ireland. The 1603 Union of the Crowns, Acts of Union 1707 and Acts of Union 1800 consolidated Britain and Ireland into a single political unit, the United Kingdom, with the Isle of Man and the Channel Islands remaining as Crown Dependencies. The expansion of the British Empire and migrations following the Irish Famine and Highland Clearances resulted in the distribution of the islands' population and culture throughout the world and a rapid de-population of Ireland in the second-half of the 19th century. Most of Ireland seceded from the United Kingdom after the Irish War of Independence and the subsequent Anglo-Irish Treaty (1919-1922), with six counties remaining in the UK as Northern Ireland. Scotland and Wales both voted for devolution within the UK in 1998, leading to the creation of a Scottish Parliament and a Welsh Assembly in 1999. Devolution of government for Northern Ireland also took place in 1999 following the Belfast Agreement of 1998, although devolution was suspended between 2002 and 2007 due to ongoing security and political problems.

The term *British Isles* is controversial in relation to Ireland, where there are objections to its usage due to the association of the word *British* with Ireland. The Government of Ireland does not use the term and its embassy in London discourages its use. As a result, *Britain and Ireland* is becoming a preferred description, and *Atlantic Archipelago* is increasingly favoured in academia, although *British Isles* is still commonly employed.

The first references to the islands as a group appeared in the writings of travellers from the ancient Greek colony of Massalia. These writings have been lost, but later writers quoted from the Massaliote Periplus (6th century BC) and from Pytheas' *On the Ocean* (circa 325–320 BC), providing several variations referring to the geographical area of the British Isles, including Britain and Ireland, which have survived. The historian Norman Davies states that "though the details are debatable, the main derivations looks pretty solid". They believe that the peoples of these islands of *Pretanike* were called the (*Priteni* or *Pretani*), names derived from a Celtic term which probably reached travellers like Pytheas from the Gauls, who may have used it as their name for the inhabitants of the islands.

The replacement of the "P" of "Pretannia" to the "B" of Britannia by the Romans occurred during the time of Julius Caesar. In classical Latin the plural term *Britannicae insulae* was sometimes used.

The earliest citation of the phrase "Brytish Iles" in the *Oxford English Dictionary* is dated 1577 in a work by John Dee.

In modern Welsh, the term *Prydain ac Iwerddon* ("Britain and Ireland") is common. *Ynysoedd Prydain* ("British Isles") usually refers to the territories of England, Scotland and Wales: it does not include Ireland, north or south.

Alternative Names and Descriptions

Several different names are currently used to describe the islands. Dictionaries, encyclopaedias and atlases that use the term *British Isles* define it as Great Britain and Ireland and adjacent islands – typically including the Isle of Man, the Hebrides, Orkney and Shetland. Some definitions include the Channel Islands. Commonly used alternative names are *British-Irish Isles*, *Britain and Ireland*, *Great Britain and Ireland*, *British Isles and Ireland*, or *UK and Ireland*.

The Government of Ireland does not use the term and its embassy in London discourages its use.

UK media organisations such as the *The Times* and the BBC have style-guide entries to try to maintain consistent usage. *Encyclopaedia Britannica*, the Oxford University Press (publishers of the *Oxford English Dictionary*) and the United

Kingdom Hydrographic Office (publisher of Admiralty charts) have all occasionally used the term *British Isles and Ireland*. The Economic History Society and the Guardian newspaper both suggest that use of the term *British Isles* should be avoided.

Some international publications no longer use the term *British Isles*. In early 2008, it was reported that *National Geographic* said it would use the wording *British and Irish Isles* instead of *British Isles*. In 2006, Folens, an Irish publisher of school text books, decided to stop using the term in Ireland while continuing to use it in the United Kingdom. Since 2001 the rugby union team the *British Isles* or *British Lions* are now named the *British and Irish Lions*.

Geography

There are about 136 permanently inhabited islands in the group, the largest two being Great Britain and Ireland. Great Britain is to the east and covers 216,777 km^2 (83,698 square miles), over half of the total landmass of the group. Ireland is to the west and covers 84,406 km^2 (32,589 square miles). The largest of the other islands are to be found in the Hebrides, Orkney and Shetland to the north, Anglesey and the Isle of Man between Great Britain and Ireland, and the Channel Islands near the coast of France.

The islands are at relatively low altitudes, with central Ireland and southern Great Britain particularly low lying: the lowest point in the islands is the Fens at "4 m ("13 ft). The Scottish Highlands in the northern part of Great Britain are mountainous, with Ben Nevis being the highest point in the British Isles at 1,344 m (4,409 ft). Other mountainous areas include Wales and parts of Ireland, but only seven peaks in these areas reach above 1,000 m (3,281 ft). Lakes on the islands are generally not large, although Lough Neagh in Northern Ireland is an exception, covering 381 km^2 (147 square miles); the largest freshwater body in Great Britain is Loch Lomond at 71.1 square kilometres (27 sq mi). There are a number of major rivers within the British Isles. The river Severn at 354 km (220 mi) is the longest in Great Britain and the Shannon at 386 km (240 mi) is the longest in Ireland.

The British Isles have a temperate marine climate; the North Atlantic Drift ("Gulf Stream") which flows from the Gulf of Mexico brings with it significant moisture and raises temperatures 11 °C (20 °F) above the global average for the islands' latitudes. Winters are cool and wet, with summers mild and also wet. Most Atlantic depressions pass to the north of the islands, combined with the general westerly circulation and interactions with the landmass, this imposes an east-west variation in climate. Geography of Ireland

The geography of Ireland describes an island in northwest Europe in the north Atlantic Ocean. The main geographical features of Ireland include low central plains surrounded by a ring of coastal mountains. The highest peak is Carrauntoohil (Irish: *Corran Tuathail*), which is 1,041 metres (3,415 ft) above sea level. The western coastline is rugged, with many islands, peninsulas, headlands and bays. The island is bisected by the River Shannon, which at 386 km (240 mi) with a 113 km (70 mi) estuary is the longest river in Ireland and flows south from County Cavan in Ulster to meet the Atlantic just south of Limerick. There are a number of sizeable lakes along Ireland's rivers, of which Lough Neagh is the largest.

Geological Development

The oldest known Irish rock is about 1.7 billion years old and is found on Inishtrahull Island off the coast of County Donegal and also at Annagh Head in north west County Mayo. In other parts of Donegal, scientists have discovered rocks that began life as glacial deposits, demonstrating that at this early period, part of what was to become Ireland was in the grip of an ice age. However, because of the effects of later upheavals, it is almost impossible to sequence these early rock layers correctly.

About 600 million years ago, at the end of the Precambrian era, the Irish landmass was divided in two, with one half on the western side of the Iapetus Ocean and the other at the eastern side, both at about the latitude that of around 80° South, close to what is now northwest Africa. From the evidence of fossils found at Bray Head in County Wicklow, Ireland was below sea level at this time.

Over the next 50 million years, these two parts drifted towards each other, eventually uniting about 440 million years ago. Fossils discovered near Clogher Head, County Louth, show the coming together of shoreline fauna from both sides of the original dividing ocean. The mountains of northwest Ireland were formed during the collision, as was the granite that is found in locations in Donegal and Wicklow. The Irish landmass was now above sea level and lying near the equator, and fossil traces of land-based life forms survive from this period. These include fossilised trees from Kiltorcan, County Kilkenny, widespread bony fish and freshwater mussel fossils and the footprints of a four-footed amphibian preserved in slate on Valentia Island, County Kerry. Old Red Sandstone also formed at this time.

Between 400 million and 300 million years ago, northwest Europe – including Ireland – sank beneath a warm, calcium-rich sea. Great coral reefs formed in these waters, eventually creating the limestone that still makes up about 65 per cent of the rock mantle of the island. As the waters receded, tropical forests and swamps flourished. The resulting vegetable debris eventually formed coal, most of which was later eroded. This period, known as the Carboniferous era, ended with further tectonic movement which saw Ireland drift further northward. The resulting pressure created those Irish mountain and hill ranges that run in a northeast to southwest direction.

By 250 million years ago, Ireland was at the latitude of present-day Egypt and had a desert climate. It was at this time that most of the coal and sandstone were eroded. The thinner layers of limestone in the south of the country were also partially affected by this erosion. The limestone that was exposed by the disappearance of its sandstone mantle was affected by carbon dioxide and other factors resulting in a karstic landscape that can still be seen in the Burren in County Clare. Shortly after this period, organic debris in the seas around Ireland began to form the natural gas and petroleum deposits that now play an important role in the economy of Ireland. Then, about 150 million years ago, Ireland was again submerged, this time in a chalky sea that resulted in the formation of chalk over large parts of the surface. Traces of this survive under the basalt lava that is found in parts of the north.

About 65 million years ago, the volcanic activity that formed this lava began. The Mourne Mountains and other mountains in the northern part of the island formed as a result of this activity. Climatic conditions at this time were warm and vegetation thrived. Vegetable debris in the Antrim depression formed deposits of brown coal or lignite which remain untouched down to the present time. The warm conditions produced high rainfall that accelerated processes of erosion and the formation of karstic landscape forms.

By 25 million years ago, Ireland was close to assuming its present position. The long period of erosion had resulted in considerable soil formation and most of the rock mantle was covered. In areas with good drainage, the covering consisted of brown or grey soil, while in poorly drained areas the black clay tended to dominate. As the climate cooled, soil formation slowed down, and a flora and fauna that would, millions of years later, be familiar to the first human inhabitants began to emerge. By about three million years ago, the present landscape of Ireland had more or less formed.

Since about 1.7 million years ago, the earth has been in the grip of a cycle of warm and cold stages and these have, inevitably, affected Ireland. The earliest evidence we have for this effect comes from the period known as the Ballylinian Warm Stage, some half a million years ago. At this time, most of what are now considered to be native Irish trees were already established on the island. The action of the ice during the cold stages was the major factor in bringing the Irish landscape to its current form.

Obvious impacts of the ice on the landscape include the formation of glacial valleys such as Glendalough in Wicklow and of corries, or glacial lakes. The depositing of mounds of debris under the melting ice created drumlins, a common feature of the landscape across the north midlands. Streams also formed under the ice and the material deposited by these formed eskers (Irish *eiscir*). The greatest of these, the Esker Riada, divides the northern and southern halves of the island and its ridge once served as the main highway connecting the east and west coasts. About one half of the coastline consists of a low lying dune pasture land known as *Machair*.

Rocks and Soil Types

The large central lowland is of limestone covered with glacial deposits of clay and sand, with widespread bogs and lakes. The Bog of Allen is one of the largest bogs. The coastal mountains vary greatly in geological structure. In the south, the mountains are composed of old red sandstone with limestone river valleys. In Galway, Mayo, Donegal, Down and Wicklow, the mountains are mainly granite, while much of the northeast of the country is a basalt plateau. An area of particular note is the Giant's Causeway, in Antrim, a mainly basalt formation caused by volcanic activity between 50–60 million years ago. The basalts were originally part of the great Thulean Plateau formed during the Paleogene period. The soils of the north and west tend to be poorly drained peats and gleys, including peaty podzols. In contrast, in the south and east the soils are free-draining brown earths and brown and grey-brown podzols. This is reflected in the rainfall distribution on the island, with the poorly drained regions being those with the highest rainfalls.

An unusual environment is present in north County Clare, in an area known as the Burren. This karst-like landscape consists of limestone bedrock, with little or no soil in the innermost areas. There are numerous sinkholes, where surface water disappears through the porous rock surface, and extensive cave systems have been formed in some areas. The Pol an Ionain cave, near Doolin, is the site of one of the world's longest known free-hanging stalactites.

Mountain Ranges

Ireland consists of a mostly flat low-lying area in the midlands, ringed by mountain ranges such as (beginning in County Kerry and working counter-clockwise) the Macgillycuddy's Reeks, Comeragh Mountains, Blackstairs Mountains, Wicklow Mountains, the Mournes, Glens of Antrim, Sperrin Mountains, Bluestack Mountains, Derryveagh Mountains, Ox Mountains, Nephinbeg Mountains and the Twelve Bens/Maumturks group. Some mountain ranges are further inland in the south of Ireland, such as the Galtee Mountains, the highest inland range, Silvermines and Slieve Bloom Mountains. The highest peak is Carrauntoohil, 1,038 m (3,405 ft) high, is in the Macgillycuddy's Reeks, a range of glacier-carved sandstone mountains in County Kerry, in the southwest of the island. The mountains are not high – only three peaks are over 1,000 m (3,281 ft) and another 457 exceed 500 m (1,640 ft).

Rivers and Lakes

The main river in Ireland is the River Shannon, 386 km (240 miles), the longest river in either Ireland or Great Britain, which separates the boggy midlands of Ireland from the West of Ireland. The river develops into three lakes along its course, Lough Allen, Lough Ree, and Lough Derg. Of these, Lough Derg is the largest. The River Shannon enters the Atlantic Ocean after Limerick city at the Shannon Estuary. Other major rivers include the River Liffey, River Lee, River Blackwater, River Nore, River Suir, River Barrow, River Erne, and River Boyne.

Lough Neagh, in Ulster, is the largest lake in Ireland and the UK. Legend has it that a giant, Fionn mac Cumhail, was fighting with another in Scotland, and enraged, scooped out a lump of earth, which he threw. It fell into the Irish Sea, creating the Isle of Man, while the hole filled up with water to become Lough Neagh. Other large lakes include Lough Erne and Lough Corrib.

Inlets

Beginning with County Donegal, Lough Swilly separates one side of the Inishowen peninsula. Lough Foyle on the other side, is one of Ireland's larger inlets, situated between County Donegal and County Londonderry. Further round the coast is Belfast Lough, between County Antrim and County Down. Also in County Down is Strangford Lough, actually an inlet partially separating the Ards peninsula from the mainland. Further down the coast, Carlingford Lough is situated between Down and County Louth.

Dublin Bay is the next sizable inlet, while the eastern coast of Ireland is mostly uniform until Wexford Harbour at the mouth of the River Slaney. On the southern coast, Waterford Harbour is situated at the mouth of the River Suir (into which the other two of the Three Sisters (River Nore and River Barrow) flow). The next major inlet is Cork Harbour, at the mouth of the River Lee, in which Great Island is situated.

Dunmanus Bay, Bantry Bay, Kenmare estuary and Dingle Bay are all inlets between the peninsulas of County Kerry. North of these is the Shannon estuary. Between north County Clare and County Galway is Galway Bay. Clew Bay is located on the coast of County Mayo, south of Achill Island, while Broadhaven Bay, Blacksod Bay and Sruth Fada Conn bays are situated on the north west part of Connaught in North Mayo. Killala Bay is on the north east coast of Mayo. Donegal Bay is a major inlet between County Donegal and County Sligo.

Headlands

Malin Head is the most northerly point in Ireland, while Mizen Head is one of the most southern points, hence the term "Malin head to Mizen head" (or the reverse) is used for anything applying to the island of Ireland as a whole. Carnsore Point is another extreme point of Ireland, being the southeasternmost point of Ireland. Further along the coast is Hook Head while the Old Head of Kinsale in is one of many headlands along the south coast of Ireland.

Loop Head is the headland at which County Clare comes to a point on the west coast of Ireland, with the Atlantic on the north, and further inland on the south, the Shannon estuary. Hag's Head is another headland further up Clare's north/western coastline, with the Cliffs of Moher along the coastline north of the point.

Erris Head is the north western most point of Connacht.

Islands and Peninsulas

Achill Island, in the northwest, is the largest island off Ireland's coast. The island is inhabited, and is connected to the mainland by a bridge. Some of the next largest islands are the Aran Islands, off the coast of southern Connacht, host to an Irish-speaking community, or Gaeltacht. Valentia Island off the Iveragh peninsula is also one of Ireland's larger islands, and is relatively settled, as well as being connected by a bridge at its southeastern end. Omey Island, off the coast of Connemara is a tidal island.

Some of the best-known peninsulas in Ireland are in County Kerry; the Dingle peninsula, the aforementioned Iveragh peninsula and the Beara peninsula. The Ards peninsula is one of the larger peninsulas outside Kerry. The Inishowen peninsula in County Donegal includes Ireland's most northerly point, Malin Head and several important towns including Buncrana on Lough Swilly, Carndonagh and Moville on Lough Foyle. Ireland's most northerly land feature is Inishtrahull island, off

Malin Head. Rockall Island may deserve this honour but its status is disputed, being claimed by the United Kingdom, Republic of Ireland, Denmark (for the Faroe Islands) and Iceland. The most southerly point is the Fastnet Rock.

Climate

Ireland's climate is temperate, though significantly warmer than almost all other locations at similar latitude, such as Poland (on the continent) or Newfoundland (on the opposite edge of the Atlantic), due to the warming influence of the North Atlantic drift. The prevailing wind blows from the southwest, breaking on the high mountains of the west coast. Rainfall is therefore a particularly prominent part of western Irish life, with Valentia Island, off the west coast of County Kerry, getting almost twice as much annual rainfall as Dublin on the east (1,400 mm/55.1 in vs. 762 mm/30.0 in). Across the country, about 60% of the annual rainfall occurs between August and January.

January and February are the coldest months of the year, and mean daily air temperatures fall between 4 and 7 °C (39.2 and 44.6 °F) during these months. July and August are the warmest, with a range of 14 to 16 °C (57.2 to 60.8 °F). The sunniest months are May and June, with an average of five to seven hours sunshine per day. Though extreme weather events in Ireland are comparatively rare when compared with other countries in the European Continent, they do occur. Explosive Atlantic depressions, occurring mainly in the months of December, January and February, can occasionally bring winds of up to 160 km/h/99 mph to Western coastal counties; while the summer months, and particularly around late July/ early August, sudden and violent thunderstorms can develop, more especially, but not exclusively, across midland and western areas of the country.

Political and Human Geography

Ireland is divided into four provinces, Connacht, Leinster, Munster and Ulster, and 32 counties. Six of the nine Ulster counties form Northern Ireland and the other 26 form the state, Ireland. The map shows the county boundaries for all 32 counties.

From an administrative viewpoint, 20 of the counties in the republic are units of local government. The other six have more than one local authority area, producing a total of 34 county-level authorities. Tipperary has two ridings, North Tipperary and South Tipperary, originally established in 1838 and renamed in 2001. The cities of Dublin, Cork, Limerick, Galway and Waterford have city councils and are administered separately from the counties bearing those names. The remaining part of County Dublin is split into Dun Laoghaire-Rathdown, Fingal, and South Dublin.

Electoral areas in Ireland (the state), called constituencies in accordance with Irish law, mostly follow county boundaries. Maintaining links to the county system is a mandatory consideration in the re-organisation of constituency boundaries..

In Northern Ireland, a major re-organisation of local government in 1973 replaced the six traditional counties and two county boroughs (Belfast and Derry) by 26 single-tier districts, which, apart from Fermanagh cross the traditional county boundaries. The six counties and two county-boroughs remain in use for purposes such as Lieutenancy. In November 2005, proposals were announced which would see the number of local authorities reduced to seven. The island's total population of approximately 6 million people is concentrated on the east coast, particularly in Dublin and Belfast and their surrounding areas.

Natural Resources

Bogs

Ireland has 12,000 km² (4,633 miles²) of bogland, consisting of two distinct types, blanket bogs and raised bogs. Blanket bogs are the more widespread of the two types. They are essentially a product of human activity aided by the moist Irish climate. Blanket bogs formed on sites where Neolithic farmers cleared trees for farming. As the land so cleared fell into disuse, the soil began to leach and become more acidic, producing a suitable environment for the growth of heather and rushes. The debris from these plants accumulated and a layer of peat formed. One of the largest expanses of Atlantic blanket bog in Ireland is to be found in Erris, County Mayo.

Raised bogs are most common in the Shannon basin. They formed when depressions left behind after the ice age filled with water to form lakes. Debris from reeds in these lakes formed a layer at the bottom of the water. This eventually choked the lakes and raised above the surface, forming raised bogs.

Since the 17th century, peat has been cut for fuel for domestic heating and cooking and it is called turf when so used. The process accelerated as commercial exploitation of bogs grew. In the 1940s, machines for cutting turf were introduced and larger-scale exploitation became possible. In the Republic, this became the responsibility of a semi-state company called Bord na Mona. In addition to domestic uses, commercially extracted turf is used in a number of industries, especially electricity generation.

In recent years, the high level of bog being destroyed by cutting has raised environmental concerns. The problem is particularly acute for raised bogs as they yield a higher-grade fuel than blanket bogs. Plans are now in place in both the Republic and Northern Ireland to conserve most of the remaining raised bogs on the island.

Oil, Natural Gas and Minerals

Offshore exploration for natural gas began in 1970. The first major discovery was the Kinsale Head gas field in 1971. Next was the smaller Ballycotton gas field in 1989, and the Corrib gas field in 1996. Exploitation of the Corrib project has yet to get off the ground because the controversial proposal to refine the gas onshore, rather than at sea has been met with widespread opposition. Gas from these fields is pumped ashore and used for both domestic and industrial purposes. The Helvick oil field, estimated to contain over 28 million barrels (4,500,000 m^3) of oil, is another recent discovery. Ireland is the largest European producer of zinc with three operating zinc-lead mines at Navan, Galmoy and Lisheen. Other mineral deposits with actual or potential commercial value include gold, silver, gypsum, talc, calcite, dolomite, roofing slate, limestone aggregate, building stone, sand and gravel.

In May 2007 the Department of Communications, Marine and Natural Resources (now replaced by the Department of Communications, Energy and Natural Resources) reported that there may be volumes over 130 billion barrels (2.1×10^{10} m^3) of petroleum and 50 trillion cubic feet (1,400 km^3) of natural gas in Irish waters-worth trillions of Euro, if true. The minimum 'guaranteed' amount of oil in the Irish Atlantic waters is 10 billion barrels (1.6×10^9 m^3), worth over €450 billion. There are also areas of petroleum and natural gas on shore, for example the Lough Allen basin, with 9.4 trillion cubic feet (270 km^3) of gas and 1.5 billion barrels (240,000,000 m^3) of oil, valued at €74.4 billion. Already some fields are being exploited, such as the Spanish Point field, with 1.25 trillion cubic feet (35 km^3) of gas and 206 million barrels (3.28E+7 m^3) of oil, valued at €19.6 billion. The Corrib Basin is also quite large, worth anything up to €87 billion, while the Dunquin gas field contains 25 trillion cubic feet (710 km^3) of natural gas and 4.13 billion barrels (657,000,000 m^3) of petroleum.

Gravity Anomalies of Britain and Ireland

Variations in the strength of gravity occur from place to place according to the density distribution of the rocks beneath the surface. Such gravity anomalies have been mapped across Great Britain and Ireland and the surrounding seas and they reveal aspects of these islands' geological structure.

Negative Anomalies

Negative gravity anomalies are associated with local deficiencies in mass due to volumes of rock which are less dense than the average. Typically these anomalies are associated with granite plutons and with sedimentary basins. Gravity anomaly maps of Britain and Ireland are interpreted as indicating the presence of granite plutons in the following areas. The list is not exhaustive:

- Galloway
- Cheviot Hills
- Mountains of Mourne
- Connemara
- Donegal
- South West England (Dartmoor, Bodmin Moor, St Austell granite etc)
- Askrigg Block (Yorkshire Dales)
- Alston Block (North Pennines)
- Grampian Highlands (Cairngorms etc).

Similarly, negative anomalies coincide with the following sedimentary basins (amongst others):

- Moray Firth Basin
- North Channel Basin
- Solway Basin
- Kish Bank Basin
- Munster Basin
- Vale of Eden
- Vale of Clwyd
- Cheshire Basin
- Worcester Graben
- Hampshire Basin
- Cardigan Bay Basin
- St. George's Channel Basin
- Wessex Basin.

Positive Anomalies

Positive gravity anomalies are associated with areas where Precambrian basement rocks are at or near the surface or where volumes of basalt or other similarly dense rocks have been put in place. Intense gravity highs coincide with a series of Palaeogene volcanic centres associated with the opening of the Atlantic Ocean and including those at:

- Skye
- St. Kilda
- Ardnamurchan
- Rum
- Dubh Artach
- Isle of Mull
- Isle of Arran
- Lundy.

In addition, intense gravity highs are associated with the Anton Dohrn and Hebrides Terrace seamounts. Gravity anomalies also accompany major lineaments, notably those associated with the Scottish Caledonides and including Great Glen Fault, Walls Boundary Fault, Highland Boundary Fault and Southern Uplands Fault.

Geography of the Isle of Man

The Isle of Man is an island in the Irish Sea, between Great Britain and Ireland in Western Europe, with a population of over 75,000. It is a British Crown dependency. It has a small islet, the Calf of Man, to its south.

Dimensions

Area:

Total: 572 km²

Land: 572 km²

Water: 0 km²

This makes it:

- slightly more than three times the size of Washington, DC
- slightly more than one third the size of Hertfordshire
- slightly smaller than Saint Lucia.

Coast

The Isle of Man has a coastline of 160 km, and claims 12 nm of territorial waters, but only holds exclusive fishing rights in the first 3 miles.

Raadny Foillan long distance footpath runs 95 miles around the Manx coast.

Climate

The Isle of Man enjoys a temperate climate, with cool summers and mild winters. Average rainfall is high compared to the majority of the British Isles, due to its location to the western side of Great Britain and sufficient distance from Ireland for moisture to be accumulated by the prevailing south-westerly winds. Average rainfall is highest at Snaefell, where it is around 1,900 mm (74.8 in) a year. At lower levels it can fall to around 800 mm (31.5 in) a year.

Temperatures remain fairly cool, with the recorded maximum being 28.9 °C (84.0 °F) at Ronalds way. The Isle of Man is known for its overcast skies and low levels of sunshine.

Terrain

The island's terrain is varied. There are two mountainous

areas divided by a central valley which runs between Douglas and Peel. The highest point in the Isle of Man, Snaefell, is in the northern area and reaches 621m. The northern end of the island is a flat plain, consisting of glacial tills and marine sediments. To the south the island is more hilly, with distinct valleys. There is no land below sea level.

Land Use

- Arable land: 9%
- Permanent crops: 0%
- Permanent pastures: 46%
- Forests and woodland: 6%
- Other: 39% (includes 25% mountain and heathland).

Natural Hazards and Environmental Issues

There are few severe natural hazards, the most common being high winds, rough seas and dense fog. In recent years there has been a marked increase in the frequency of high winds, heavy rains, summer droughts and flooding both from heavy rain and from high seas. Snow fall has decreased significantly over the past century while temperatures are increasing year round with rainfall decreasing.

Air pollution, marine pollution and waste disposal are issues in the Isle of Man.

Global warming and a sea level rise potentially pose a great threat to the Isle of Man. All of the Island's towns are at threat from rising sea levels while the Northern Plain, a large, flat and low-lying plain composed of soft marine sediments and glacial material, which makes up about a quarter of the Island's landmass, is in danger of being lost to the sea over the next two centuries. The same is true for the considerably smaller Southern Plain surrounding the settlements of Castletown and Ballasalla and including Ronalds way Airport.

Geography of the United Kingdom

The United Kingdom of Great Britain and Northern Ireland, or UK. is a sovereign state located off the northwestern coast of continental Europe. It comprises the island of Great Britain (England, Scotland and Wales) and the northeastern one-sixth of the island of Ireland (Northern Ireland), together with many smaller islands. The mainland areas lie between latitudes 49°N and 59°N (the Shetland Islands reach to nearly 61°N), and longitudes 8°W to 2°E. The Royal Greenwich Observatory, in South East London, is the defining point of the Prime Meridian.

The UK lies between the North Atlantic and the North Sea, and comes within 35 km (22 mi) of the northwest coast of France, from which it is separated by the English Channel. Northern Ireland shares a 360 km international land boundary with the Republic of Ireland. The Channel Tunnel ("Chunnel") bored beneath the English Channel, now links the UK with France.

Area

The total area of the United Kingdom is approximately 245,000 square kilometres (94,600 sq mi) comprising of the island of Great Britain, the northeastern one-sixth of the island of Ireland (Northern Ireland) and smaller islands. England is the largest country of the United Kingdom, at 130,410 square kilometres (50,350 sq mi) accounting for just over half the total area of the UK. Scotland at 78,772 square kilometres (30,410 sq mi), is second largest, accounting for about a third of the area of the UK. Wales and Northern Ireland are much smaller, covering 20,758 square kilometres (8,010 sq mi) and 14,160 square kilometres (5,470 sq mi) respectively.

The area of the countries of the United Kingdom is set out in the table below. Information about the area of England, the largest country, is also broken down by region.

Rank	*Name*	*Area*
1	England	130,410 km^2
2	Scotland	78,772 km^2
3	Wales	20,758 km^2
4	Northern Ireland	14,160 km^2
5	United Kingdom	244,100 km^2

Physical Geography

The physical geography of the UK varies greatly. The Geography of England consists of lowland terrain, with mountainous terrain north-west of the Tees-Exe line including the Cumbrian Mountains of the Lake District, the Pennines and limestone hills of the Peak District, Exmoor and Dartmoor.

The Geography of Scotland is distinguished by the Highland Boundary Fault – a geological rock fracture – which traverses the Scottish mainland from Helensburgh to Stonehaven. The faultline separates the two distinctively different regions of the Highlands to the north and west and the lowlands to the south and east. The Geography of Wales is mostly mountainous, though south Wales is less mountainous than north and mid Wales. The Geography of Ireland includes the Mourne Mountains as well as Lough Neagh, at 388 square kilometres (150 sq mi), the largest body of water in the UK and Ireland.

The overall geomorphology of the UK was shaped by the combined forces of tectonics and climate change, in particular glaciation.

The exact centre of the island of Great Britain is disputed. Depending upon how it is calculated it can be either Haltwhistle in Northumberland, or Dunsop Bridge in Lancashire.

Geology

The geology of the United Kingdom is varied and diverse. This gives up to the wide variety of landscapes found across the UK. This variety, coupled with the early efforts of UK based scientists and geologists to understand it, has influenced the naming of many geological concepts, including many of the geological periods (for example, the Ordovician period is named after the Ordovices, a people of early Britain; the Devonian period is named after the county of Devon in south-west England).

The oldest rocks in the UK are gneisses which date from at least 2,700 Ma ("Ma" means "millions of years ago") in the Archaean Period, which are found in the far north west of Scotland and in the Hebrides, with a few small outcrops elsewhere. South of the gneisses are a complex mixture of rocks forming the North West Highlands and Grampian Highlands in Scotland. These are essentially the remains of folded sedimentary rock, deposited over the gneiss, from 1,000 Ma, with a notable 7 km thick layer of Torridon Sandstone being deposited about 800 Ma, as well as the debris deposited by an ice sheet 670 Ma.

The remains of ancient volcanic islands underlie much of central England with small outcrops visible in many places.

Around 600 Ma, the Cadomian Orogeny (mountain building period) caused the English and Welsh landscape to be transformed into a mountainous region, along with much of north west Europe.

The Welsh Skiddaw slate deposits formed at around 500 Ma, during the Ordovician Period. At about this time, around 425 Ma, north Wales (and south Mayo in Ireland) experienced volcanic activity. The remains of these volcanoes are still visible, for example Rhobell Fwar, dating from 510 Ma. Large quantities of volcanic lava and ash known as the Borrowdale Volcanics covered both Wales and the Lake District, still seen in the form of mountains such as Helvellyn and Scafell Pike.

In the Silurian Period, between 425 and 400 Ma, the Caledonian fold mountains formed (the Caledonian Orogeny), covering much of what is now the UK to perhaps 8,000 feet (2,438 m) thick. Volcanic ashes and lavas deposited during this period are still found in the Mendip Hills and in Pembrokeshire.

Volcanic deposits formed Ben Nevis in the Devonian Period. Sea levels varied considerably, with the coastline advancing and retreating from north to south across England, and with the deposition of numerous sedimentary rock layers. The Old Red Sandstone of Devon gave the period its name, though deposits are found in many other places.

During the Carboniferous Period, around 360 Ma, the UK was lying at the equator, covered by the warm shallow waters of the Rheic Ocean, during which time the Carboniferous limestone was deposited, still found in areas such as the Mendip Hills and the Pennines. The coal measures were formed at this time, in river deltas, swamps and rain forests. Coal can be found in many areas of the UK, as far North as Sutherland and as far south as Kent, though it has largely been mined in the Midlands, northern England and Wales. Also formed were the Millstone Grits.

During the Permian and Triassic Periods, much of the UK was beneath shallow seas, leading to the deposition of sedimentary rocks such as shale, limestone, gravel, and marl. The seas finally receded to leave a flat desert with salt pans.

At the beginning of the Jurassic Period, the UK was underwater again, leading to the deposition of sedimentary rocks which now underlie much of England from the Cleveland Hills of Yorkshire to the Jurassic Coast in Dorset, including clays, sandstones, and the oolitic limestone of the Cotswold Hills. The burial of algae and bacteria below the mud of the sea floor during this time resulted in the formation of North Sea oil and natural gas. In the Cretaceous Period, much of the UK was again below the sea and chalk and flints were deposited over much of Great Britain. These are now notably exposed at the White Cliffs of Dover, and form Salisbury Plain, the Chiltern Hills, the South Downs and other similar features.

The last volcanic rocks in the UK were formed in the early Tertiary Period, between 63 and 52 Ma, with the major eruptions that formed the Antrim Plateau and the basaltic columns of the Giant's Causeway. Further sediments were deposited over southern England, including the London clay, while the English Channel consisted of mud flats and river deposited sands.

The major changes during the last few million years, during the Quaternary Period, have been brought about by several recent ice ages, leaving a legacy of U-shaped valleys in highland areas, and fertile (if often stoney) soil in southern U.K.

Mountains and Hills

The ten tallest mountains in the UK are all found in Scotland. The highest peaks in each part of the UK are:

- Scotland: Ben Nevis (Aonach Mor, 1,344 metres)
- Wales: Snowdon (Snowdonia, 1,085 metres)
- England: Scafell Pike (Cumbrian Mountains, 977 metres)
- Northern Ireland: Slieve Donard (Mourne Mountains, 852 metres).

The ranges of mountains and hills in the UK include:

- Scotland: Cairngorms, Cheviot Hills, Scottish Highlands, Southern Uplands, Grampian Mountains
- Wales: Brecon Beacons, Cambrian Mountains, Snowdonia, Black Mountains, Preseli Hills
- England: Chilterns, Cotswolds, Dartmoor, Lincolnshire Wolds, Exmoor, Lake District, Malvern Hills, Mendip Hills, North Downs, Peak District, Pennines, Salisbury Plain, South Downs, Shropshire Hills, Yorkshire Wolds

- Northern Ireland: Mourne Mountains, Antrim Plateau, Sperrin Mountains.

The lowest point of the UK is in the Fens of East Anglia, in England, parts of which lie up to 4 metres below sea level.

Rivers and Lakes

- List of lakes in the United Kingdom;
- Rivers of the United Kingdom;
- Waterfalls of the United Kingdom.

The longest river in the UK is the River Severn (220 mi, 354 km) which flows through both Wales and England.

The longest rivers in the UK by country are:

- England: River Thames (215 mi, 346 km)
- Scotland: River Tay (117 mi, 188 km)
- N. Ireland: River Bann (76 mi, 122 km)
- Wales: River Tywi (64 mi, 103 km).

The largest lakes in the UK by country are:

- N. Ireland: Lough Neagh (147.39 sq mi, 381.74 km^2)
- Scotland: Loch Lomond (27.46 sq mi, 71.12 km^2)
- England: Windermere (5.69 sq mi, 14.74 km^2)
- Wales: Llyn Tegid (Bala Lake) (1.87 sq mi, 4.84 km^2).

The deepest lake in the UK is Loch Morar with a maximum depth of 309 metres (Loch Ness is second at 228 metres deep). The deepest lake in England is Wastwater which descends to 79 metres (258 feet).

Artificial Waterways

As a result of its industrial history, the United Kingdom has an extensive system of canals, mostly built in the early years of the Industrial Revolution, before the rise of competition from the railways. The United Kingdom also has numerous dams and reservoirs to store water for drinking and industry. The generation of hydroelectric power is rather limited, supplying less than 2% of British electricity mainly from the Scottish Highlands.

Coastline

The UK has a coastline which measures about 12,429 km. The heavy indentation of the coastline helps to ensure that no location is more than 125 km from tidal waters. The UK claims jurisdiction over the continental shelf, as defined in continental shelf orders or in accordance with agreed upon boundaries, an exclusive fishing zone of 200 nmi (370.4 km; 230.2 mi), and territorial sea of 12 nmi (22.2 km; 13.8 mi).

Inlets

- Carbis Bay
- Cardigan Bay
- Lyme Bay
- Bristol Channel
- Thames estuary
- Morecambe Bay
- Solway Firth
- The Wash
- Humber estuary
- Firth of Forth
- Firth of Tay
- Moray Firth.

Headlands

This list is incomplete; you can help by expanding it.

The geology of the UK is such that there are many headlands along its coast, here are some of the most notable ones:

- Corwall
  - Land's End
  - The Lizard
  - Cape Cornwall
- Cumbria
  - Furness
- Devon
  - Start Point
  - Berry Head
  - Hartland Point
  - Foreland Point
- Dorset
  - Old Harry Rocks

- St. Alban's Head
- Portland Bill

- Isle of Wight
  - The Needles
  - St. Catherine's Point
- Kent
  - North Foreland
- West Sussex
  - Selsey Bill
- East Sussex
  - Beachy Head
- Sutherland
  - Cape Wrath
- Wales
  - Gower Peninsula
  - Lleyn Peninsula
- Wigtownshire
  - Rhins of Galloway
- Yorkshire
  - Filey Brigg
  - Flamborough Head
  - Spurn Head.

Coastal Change

Islands

In total, it is estimated that the UK is made up of over one thousand small islands, some being natural and some being man-made crannogs, which were built in past times using stone and wood and which were enlarged by natural waste building up over time.

- Islands of England;
  - Lundy
  - Isles of Scilly
  - Isle of Wight
  - Farne Islands
  - Lindisfarne

  - o Isle of Portland
  - o Walney Island
  - o Burgh Island
- Islands of Scotland;
  - o Orkney Islands
  - o Shetland Islands
    - * Fair Isle
  - o Inner Hebrides
  - o Outer Hebrides
  - o Rockall
  - o Bass Rock.
- Islands of Wales;
  - o Anglesey
  - o Skomer Island
  - o Skokholm Island
  - o Ramsey Island
  - o Bardsey Island
  - o Holy Island.
- Islands of Northern Ireland;
  - o Rathlin Island.

Climate

The climate of the UK varies, but is generally temperate, though significantly warmer than some other locations at similar latitude, such as central Poland, due to the warming influence of the Gulf Stream. In general, the south is warmer and drier than the north.

The prevailing winds are southwesterly, from the North Atlantic Current. More than 50% of the days are overcast. There are few natural hazards, although there can be strong winds and floods, especially in winter.

Average annual rainfall varies from over 3,000 mm (118.1 in) in the Scottish Highlands down to 553 mm (21.8 in) in Cambridge. The county of Essex is one of the driest in the UK, with an average annual rainfall of around 600 mm (23.6 in), although it typically rains on over 100 days per year. In some years rainfall in Essex can be below 450 mm (17.7 in), less than the average annual rainfall in Jerusalem and Beirut. The highest temperature recorded in the UK was 38.5 °C (101.3 °F) at Brogdale, near Faversham, in the county of Kent, on 10 August 2003. The lowest was “27.2 °C (“17 °F) recorded at Braemar in the Grampian Mountains, Scotland, on 11 February 1895 and 10 January 1982 and Altnaharra, also in Scotland, on 30 December 1995.

Human Geography

Demographics

According to the 2001 census, the total population of the United Kingdom was 58,789,194—the third-largest in the European Union (behind Germany and metropolitan France) and the 21st-largest in the world. Its overall population density is one of the highest in the world, due to the particularly high population density in England. Almost one-third of the population lives in England's southeast and is predominantly urban and suburban, with about 8.2 million in the capital city of England, London. The United Kingdom's extremely high literacy rate (99%) is attributable to universal public education introduced for the primary level in 1870 (Scotland 1872, free 1890) and secondary level in 1900. Parents are obliged to have their children educated from the ages of 5 to 16 (with legislation passed to raise this to 18), and can continue education free of charge in the form of A-Levels, vocational training or apprenticeship to age 18. About one-fifth of British students go on to post-secondary education (18+). The Church of England and the Church of Scotland function as the national churches in their respective countries, but all the major religions found in the world are represented in the United Kingdom.

The UK's population is predominantly White British. Being located close to continental Europe, the countries that formed the United Kingdom were subject to many invasions and migrations, especially from Scandinavia and the continent, including Roman occupation for several centuries. Historically, British people were thought to be descended mainly from the different ethnic stocks that settled there before the 11th century; The Britons, Anglo-Saxons, Romans, Vikings and the Normans, who had lived in Northern France. However, recent genetic analysis appears to indicate that around 80% of British DNA comes from an indigenous population who settled Britain around 12,000 years ago, with subsequent invaders contributing very little to the genepool. Although Celtic languages are partially spoken in Scotland, Cornwall, and Northern Ireland, the predominant language overall is English. In North and West Wales, Welsh is widely spoken as a first language, but much less so in the more English dominated South East of the country.

History

Three sets of demographic statistics are useful to governments and others concerned with their nations' political and economic stability. The first is an enumeration of the number of inhabitants distinguished by age, sex, and occupation. The second involves a continuous record of population trends from the registration of births, marriages, and burials. The third is documentation of the extent of internal and external migration.

England and Wales

Before 1800, England had none of these except for the civil registration of births, marriages, and burials briefly attempted under the Commonwealth (1653–1660) and an even more short-lived initiative of the same kind in 1694 in connection with the attempt to raise a tax on the occasion of every birth, marriage, and death—paupers excepted. At that time, the chief source of information on the demography was provided by parish registration of baptisms, marriages, and burials that had occurred in the parish churches, supplemented by information on mortality in the Bills of Mortality that were published for certain large towns and by inferences drawn from various counts of taxpayers. The article focuses on the reliability of the parochial registration system and the way in which it was exploited by the state as measured against the state's objectives for establishing it in 1538. These objectives were rarely achieved. By the end of the 18th century, the parish registers were falling short of providing a national system of registration. Neither had the registers at any time provided the requisite detail to allow the verification of age, lineal descent, or right of inheritance. They had not been used as a way of raising revenue except briefly between 1694 and 1705. Moreover, the Anglican Church was extremely lax about the enforcement of its own regulations regarding the appropriate time for registering baptisms, burials, and marriages.

The ability of the registration system to fulfil these original objectives can be measured in terms of the breadth of its coverage and the quality of the information provided. Each category can be further subdivided. For example, the breadth of coverage can be defined to include the speed with which parishes throughout the country commenced the registration of baptisms, marriages, and burials; the percentage of the population whose vital events went unrecorded even in the parishes that established registers; and the success of the incumbents and churchwardens in preserving the registers completed by their predecessors. The quality of the recording can be assessed based on the amount of information offered about individuals mentioned in the registers, the extent to which that information was provided in a standard form across the country, and the clarity of the presentation (whether separate registers for baptisms, marriages, and burials were maintained). The accuracy of the work undertaken by the parochial clergy as unpaid servants of the state in providing Rickman with the totals of baptisms, marriages, and burials can also be assessed. Each of these aspects will now be considered in turn before an assessment of the overall effectiveness of the registration system is attempted. Cromwell's brief instructions establishing the registration system did not specify what sort of register book was to be provided. Not unnaturally, most parishes chose to use paper rather than the dearer, but more durable, parchment. In 1597, the Convocation of the clergy, bishops, and archbishop of the province of Canterbury found it necessary to order parchment copies of all entries from old paper registers, instructions that were soon reissued in 1603.

The first census held throughout the UK was organised in 1801. England and Wales started the civil registration of births, marriages, and deaths in 1837. The first attempt by the state to compile statistics on migration was included in the census of 1841.

United Kingdom

During the Industrial Revolution, the life expectancy of children increased dramatically. The percentage of the children born in London who died before the age of five decreased from 74.5% in 1730–1749 to 31.8% in 1810–1829. According to Robert Hughes in *The Fatal Shore*, the population of England and Wales, which had remained steady at 6 million from 1700 to 1740, rose dramatically after 1740.

The first Census in 1801 revealed that the population of England, Scotland and Wales was 10.5 million. In 1800 the population of Ireland was between 4.5 and 5.5 million.

The 1841 UK Census counted the population of England and Wales to be 15.9 million. Ireland's population was 8.2 million in 1841. The population of Scotland was 2.6 million.

The Great Irish Famine, which began in the 1840s, caused the deaths of one million Irish people, and caused over a million to emigrate. Mass emigration became entrenched as a result of the famine and the population continued to decline until late in the 19th century.

The population of England had almost doubled from 16.8 million in 1851 to 30.5 million in 1901. Ireland's population decreased rapidly, from 8.2 million in 1841 to less than 4.5 million in 1901.

Population

The population of the United Kingdom in the 2001 census was 58,789,194. This has since risen to 60,587,300 according to mid-2006 estimates by the Office for National Statistics.

Age Structure

The most numerous age groups (at the 2001 census) were the 5-year group born in the years 1946-51 (the post-World War II baby boom); the baby boom born a generation later in 1966-71 (the largest group of all); and a more modest boom a generation after that, born in 1986-91. The 1946-51 group reaches retirement age from 2006 onwards (women from 2006 and men from 2011), and the sudden increase in the number of people claiming the state pension has led politicians and political commentators to fear a "pensions crisis".

The age groups are not evenly distributed around the country, with some areas having many young adults and children and some areas having large numbers of older people.

Fertility

In 2008 the UK's total fertility rate (TFR) was 1.96 children per woman, below the replacement rate of 2.10. In 2001, the TFR was at a record low of 1.63, but it has increased each year since. The TFR was considerably higher during the 1960s 'baby boom', peaking at 2.95 children per woman in 1964. TFR varies widely across the UK, standing at 2.10 in Northern Ireland, 1.80 in Scotland, 1.94 in England and 1.96 in Wales in 2008.

The TFR for British residents also varies by country of birth. In England and Wales in 1996, people born in the UK had a TFR of 1.67, India 2.21 and Pakistan and Bangladesh 4.90, for example.

Most of the increase in overall fertility in England and Wales can be attributed to rising fertility among UK born women, who make up the majority of the female population of child bearing age (85 per cent in 2007). According to new estimates, the TFR for women born in the UK has risen substantially, from 1.68 in 2004 to 1.79 in 2007.

Religion

The traditional religion in the United Kingdom is Christianity. In England the established church is the Church of England (Anglican). In Scotland, the Church of Scotland (a Presbyterian Church) is regarded as the 'national church' but there is not an established church. In Wales there is no established church, with the Church in Wales having been disestablished in 1920. Likewise, in Northern Ireland the Church of Ireland was disestablished in 1871. In Northern Ireland and similarly in parts of Scotland, there is a sectarian divide between Roman Catholic and Protestant communities.

The table below shows the most recent census data (2001) regarding religious belief:

Religion	Number	Percentage
Christian	42,079,000	71.6%
Muslim	1,591,000	2.7%
Hindu	559,000	1.0%
Sikh	336,000	0.6%
Jewish	267,000	0.5%
Buddhist	152,000	0.3%

Other religion	· 179,000	0.3%
Not stated	4,289,000	7.3%
No religion	9,104,000	15.5%
Total religious	45,163,000	76.8%

Rather than select one of the specified religions offered on the 2001 Census form, many people chose to write in their own religion. Some of these religions were reassigned to one of the main religions offered, predominantly within the Christian group.

In England and Wales, 151,000 people belonged to religious groups which did not fall into any of the main religions. The largest of these were Spiritualists (32,000) and Pagans (31,000), followed by Jain (15,000), Wicca (7,000), Rastafarian (5,000), Bahai (5,000) and Zoroastrian (4,000).

Although the Census 2001 also recorded 390,000 Jedi Knights, making Jedi the fourth-largest "religion" in the UK, this does not confer them any official recognition. In fact, all returns with "Jedi Knight" were classified as "No religion", along with Atheist, Agnostic, Heathen and those who ticked "Other" but did not write in any religion.

See also: Status of religious freedom in the United Kingdom, Islam in the United Kingdom, Hinduism in the United Kingdom, Buddhism in the United Kingdom, Sikhism in the United Kingdom, Judaism in the United Kingdom

Languages

The United Kingdom has no official language. The dominant language, spoken as a first language by 95% of the population, is English. Scots is spoken by around 500,000 people in Scotland and 30,000 in Northern Ireland, where it is called Ulster Scots. Welsh is spoken by around 600,000 people. Scottish Gaelic is spoken by about 60,000 speakers, mostly in Scotland. Cornish is spoken by around 2,500 people.

Irish is spoken by about 106,844 speakers in Northern Ireland. The Polish minority in the United Kingdom estimated over 600,000 people speak mostly Polish at home, Polish Britons are mainly Polish-born immigrants to the UK, although many are those who settled in Britain after the second world war and their descendants.

Education

Each country of the United Kingdom has a separate education system, with power over education matters in Scotland, Wales and Northern Ireland being devolved.

The Secretary of State for Children, Schools and Families and the Secretary of State for Innovation, Universities and Skills are responsible to the UK Parliament for education in England, though the day to day administration and funding of state schools is the responsibility of Local Education Authorities. Universal state education in England and Wales was introduced for primary level in 1870 and secondary level in 1900. Education is mandatory from ages five to sixteen (15 if born in late July or August). The majority of children are educated in state-sector schools, only a small proportion of which select on the grounds of academic ability. Despite a fall in actual numbers, the proportion of children in England attending private schools has risen to over 7%. Just over half of students at the leading universities of Cambridge and Oxford had attended state schools. State schools which are allowed to select pupils according to intelligence and academic ability can achieve comparable results to the most selective private schools: out of the top ten performing schools in terms of GCSE results in 2006 two were state-run grammar schools. England has some of the top universities in the world with Cambridge, Oxford, and London ranked amongst the top 20 in the 2007 THES-QS World University Rankings.

In Scotland, the Cabinet Secretary for Education and Lifelong Learning is responsible to the Scottish Parliament for education, with day to day administration and funding of state schools being the responsibility of Local Authorities. Scotland first legislated for universal provision of education in 1696. The proportion of children in Scotland attending private schools is just over 4% though it has been rising slowly in recent years. Scottish students who attend Scottish universities pay neither tuition fees nor graduate endowment charges as the fees were abolished in 2001 and the graduate endowment scheme was abolished in 2008.

The National Assembly for Wales has responsibility for education in Wales. A significant number of students in Wales are educated either wholly or largely through the medium of

Welsh and lessons in the language are compulsory for all until the age of 16. There are plans to increase the provision of Welsh Medium schools as part of the policy of having a fully bi-lingual Wales.

The Northern Ireland Assembly is responsible for education in Northern Ireland though responsibility at a local level is administered by 5 Education and Library Boards covering different geographical areas.

Political Geography

National Government

The UK is governed as a whole by the Parliament of the United Kingdom.

Of the four countries that make the UK, Scotland, Wales and Northern Ireland have devolved administrations and parliament/assembly:

- Northern Ireland-Northern Ireland Assembly
- Scotland-Scottish Parliament
- Wales-Welsh Assembly.

England has no devolved system of government and is governed by UK government ministers and legislated for by the UK parliament. Within England, London has a devolved assembly but proposals for elected Regional Assemblies in England were rejected in the first referendum covering North East England. *See Government of England.*

The UK (specifically, Northern Ireland) has an international land boundary with the Republic of Ireland of 360 km. There is also a boundary between the jurisdiction of France and of the UK on the Channel Tunnel.

Local Government

Each part of the UK is subdivided in further local governmental regions:

- England: Unitary Authorities, county councils, district councils, parish councils
- Wales: Principal areas, communities
- Scotland: Council areas, communities
- Northern Ireland: Districts.

Historically the UK was divided into counties or shires: administrative areas through which all civil responsibilities of the government were passed. Each county or shire had a county town as its administrative centre and was divided into individual parishes that were defined along ecclesiastic boundaries.

Between 1889 (1890 in Scotland) and 1974, the political boundaries were based on the traditional counties, but due to changes in population centres, the traditional counties became impractical as local government areas in certain highly urbanised areas. The Local Government Act 1972 created a new system of administrative counties, designed to take account of the widely differing populations across different parts of the country.

In the 1990s further population growth led to more political changes on a local level. Unitary authorities were formed across the entire of Scotland and Wales, and in larger cities in England. Many unpopular administrative counties were also abolished at this time, leading to a mixture of two-tier and single-purpose authorities. Further reorganisations are planned if and when regional assemblies in England are revisited in the future.

Economic Geography

The economic geography of the United Kingdom reflects its high position in the current economic league tables, as well as reflecting its long history as a trading nation and as an imperial power. This in turn was built on exploitation of natural resources such as coal and iron ore.

Much has changed since Bevan's speech in 1945, with the coalfields largely deserted and the Empire relinquished. With its dominant position gone, the UK economic geography is increasingly shaped by the one constant: it is a trading nation.

Agriculture

The UK has rarely been self-sufficient in terms of food supply and its modern pattern of agriculture reflects a combination of history, current public policy and comparative advantage.

Since the Enclosure Acts of the eighteenth century, the UK's uplands (including Wales and the Scottish Highlands) have largely been associated with animal husbandry and forestry. However, by the time of the Enclosure Acts, most of lowland Britain was already enclosed by processes such as assarting or illegal, but tolerated, piecemeal enclosure. However, evidence of the former open field system of agriculture can still be seen in some parts of the landscape, such as in the indentations remaining from boundary ditches of the former farming strips. Enclosure, in turn, led to intensification.

Most UK agriculture is intensive and highly mechanised, with the use of chemical fertilisers and insecticides routine. By European standards it is very efficient, although that does not necessarily make it profitable. This intense nature was compounded in the post-War years, with fields being expanded at the expense of hedgrows. This process has been heavily criticised for damaging biodiversity.

East Anglia and South East England have been centres for grain production, with some areas of South East England also specialising in market gardening. The county of Kent was so well-known for this that it is often referred to as the *Garden of England* and was particularly noted for hop growing. Dairy farming is most prevalent in South West England.

The detailed pattern of modern UK agriculture is heavily influenced by the Common Agricultural Policy of the European Union, with a combination of price support and set-aside policy.

Around the edges of south eastern towns, perfectly good agricultural land often remains uncultivated as a result of price distortions created by the Metropolitan Green Belt. Indeed, since around 2001 speculators have been buying Green Belt agricultural land, generally adjacent to built up areas, and selling it off in plots, persuading buyers that the government will have to weaken Green Belt protection to solve the housing crisis. It is hard to see when such plots could come back into production.

Primary Industry

The UK's primary industry sector was once dominated by the coal industry, heavily concentrated in south Wales, Midlands, Yorkshire, North East England and southern Scotland. The number of pits and miners have been slashed, and output fell by more than 75% between 1981 and 2003. The remaining pits produced 17.2 million tonnes of oil equivalent in 2003, making the UK the 15th largest coal producing nation, compared with 4th in 1981, according to the BP Statistical Review of World Energy 2004.

The major primary industry is North Sea oil. Its activity is concentrated on the east coast of Scotland and North East England. The waters in the North Sea off the east coast of Scotland contain nearly half of the UK's remaining oil reserves, and a quarter of reserves are located in the North Sea near the Shetland Islands. As of January 2004, the UK had proven crude oil reserves of 4.7 billion barrels (750,000,000 m^3) including onshore reserves, according to the *Oil and Gas Journal*.

A closely related industry is natural gas which, since the 1970s, has supplied all of the UK gas needs, replacing poisonous coal, or town, gas. Most natural gas production is in the North Sea, with a small amount onshore and in the Irish Sea. The largest reserves not related to oil production are in the southern North Sea between the UK and the Netherlands, although; the largest reserves, are associated with oil production. The UK is expected to become a net natural gas importer before 2010.

Manufacturing

At one time or another virtually every product that can be imagined has been made in the UK. In particular its heavy manufacturing drove the industrial revolution, starting with the first blast furnace at Coalbrookdale in Shropshire.

A map of the major UK cities gives a good picture of where manufacturing flourished, and often specialisations could be identified, in particular:

- Birmingham, (automotive)
- Glasgow (shipbuilding),
- London (everything),
- Manchester (textiles),
- Newcastle (shipbuilding and steel),
- Nottingham (apparel, medicine),
- Sheffield (steel and steel products).
- Leeds (textiles and engineering)
- Belfast (shipbuilding and textiles)
- Cardiff (steel).

In the inter-War years modern industries emerged, with aerospace forming clusters around Bristol and in Hertfordshire. The Hertfordshire cluster no longer exists. The early electronics industry generally preferred the south, especially the home counties.

Today there is no heavy manufacturing industry in which UK-based firms can be considered world leaders and no product in which a UK city or region is *the* world leader.

However, the Midlands, in particular, remains a strong manufacturing centre, with around a fifth of employment dependent on manufacturing, and the East Midlands Development Agency has a policy to maintaining this characteristic.

More recently, high technology firms have concentrated largely along the M4 motorway, partly because of access to Heathrow Airport, but also because of agglomeration economies. But the general pattern remains that south has lower, and falling, reliance on manufacturing.

Finance and Services

Once, every great city had a stock exchange. Now, the UK financial industry is concentrated overwhelmingly in the City of London and Canary Wharf, with back office and administrative operation often dispersed around the south of England.

London is one of the world's great financial centres, which is one of the factors that is commonly considered to make it a world city. Central London contains some of the most expensive commercial property in the world because of this.

From around the early 1990s London has been able to boast of having more U.S. banks than New York, as well as being host to branches of more than five hundred overseas banks. It is the principal financial centre of the world, ranked alongside New York and Tokyo as one of three where any serious financial player *must* be represented.

The City of London has around 300,000 employees, largely concentrated in the financial and professional sectors.

Within London, the desire for banks to be there put pressure on the City of London's ability to accommodate them. In the mid-1980s a crisis point was reached and, although the City was able to expand its stock of modern office space, this did not happen before Canary Wharf had been set up as a competitor. The irony is that Canary Wharf was built in the London Docklands that had, until 1970, been among the largest docks in the world. All manner of goods had been shipped from the factories of east London to the world.

During the mid-90's, Leeds took advantage of the internet's arrival as part of the.com boom. It became the UK's first city to have full broadband and digital coverage, making it a very attractive place for corporations to expand, particularly when opening call centres. During the changing 1990's, Leeds had better connections to London than any other UK city, being 2 hours from London by train. As a result, it is now the UK's 2nd largest financial and legal centre, and the UK's largest e-business sector with more than 1/3rd of the UK's internet traffic passing through the city.

In this respect, Canary Wharf and Leeds are arguably more symbolic of the changed economic geography of the UK than any other place.

Regional Disparity

The combined effects of changing economic fortunes, economic restructuring and the decline of the UK as an imperial power have created the so-called North-South divide, in which decaying industrial areas of the north of England and Scotland contrast with the wealthy, finance-and-technology led southern economy. This has led successive governments to develop regional policies to try and rectify the imbalance.

The success has been, at best, patchy and the uneven distribution of economic wealth in the UK has led to migration from north to south, aggravating serious pressure on the southern housing market.

Although, in 2004, house prices in the north of England and Scotland increased faster than those in the south, this happened at the same stage of the last property cycle and the rises are off of a lower value base-it is far too early to concluded whether the trend is reversing. Either way, there were 113,000 transactions in London and the South East recorded by the UK HM Land Registry in the year to November 2004, compared with 83,000 in the North, North West and Yorkshire and the Humber.

This is not to say that the south is uniformly wealthy; Some of the worst pockets of deprivation can be found in London, especially inner London, while Hastings and Bexhill-on-Sea, on the south coast are also subject to urgent regeneration efforts.

Natural Resources

Historically, much of the United Kingdom was forested. Since prehistoric times, man has deforested much of the United Kingdom. Agriculture is intensive, highly mechanised, and efficient by European standards, producing about 60% of food needs with only 1% of the labour force. It contributes around 2% of GDP. Around two thirds of production is devoted to livestock, one third to arable crops.

In 1993, it was estimated that land use was:

- Arable land: 25 %
- Permanent crops: 0 %
- Permanent pastures: 46 %
- Forests and Woodland: 10 %
- Other: 19 %
- Irrigated: 1,080 km^2

The UK has a variety of natural resources including:

- Geological: coal, petroleum, natural gas, limestone, chalk, gypsum, silica, rock salt, china clay, iron ore, tin, silver, gold, lead.
- Agricultural: arable land, wheat, barley, sheep.

The UK has large coal, natural gas, and oil reserves; primary energy production accounts for 10% of GDP, one of the highest shares of any industrial nation. Due to the island location of the UK, the country has great potential for generating electricity from wave power and tidal power, although these have not yet been exploited on a commercial basis.

Environment

Current Issues

The United Kingdom is reducing greenhouse gas emissions. It has met Kyoto Protocol target of a 12.5 % reduction from

1990 levels and intends to meet the legally binding target of a 20 % cut in emissions by 2010. By 2015, to recycle or compost at least 33 % of household waste. Between 1998-99 and 1999–2000, household recycling increased from 8.8 % to 10.3 % respectively.

International Agreements

The United Kingdom is a party to many international agreements, including: Air Pollution, Air Pollution-Nitrogen Oxides, Air Pollution-Sulphur 94, Air Pollution-Volatile Organic Compounds, Antarctic-Environmental Protocol, Antarctic-Marine Living Resources, Antarctic Seals, Antarctic Treaty, Biodiversity, Climate Change, Climate Change-Kyoto Protocol, Desertification, Endangered Species, Environmental Modification, Hazardous Wastes, Law of the Sea, Marine Dumping, Marine Life Conservation, Nuclear Test Ban, Ozone Layer Protection, Ship Pollution, Tropical Timber 83, Tropical Timber 94, Wetlands and Whaling.

The UK has signed, but not ratified, the international agreement on Air Pollution-Persistent Organic Pollutants.

Channel Islands Geography

The Channel Islands are an archipelago of British Crown Dependencies in the English Channel, off the French coast of Normandy. They include two separate bailiwicks: the Bailiwick of Guernsey and the Bailiwick of Jersey, neither of which is part of the United Kingdom; rather they are considered the remnants of the Duchy of Normandy. They have a total population of about 158,000 and their respective capitals, St. Peter Port and St. Helier, have populations of 16,488 and 28,310. The total area of the islands is 194 km^2.

The Bailiwicks have been administered separately from each other since the late 13th century, and although those unacquainted with the islands often assume they form one political unit, common institutions are the exception rather than the rule. The two Bailiwicks have no common laws, no common elections, and no common representative body (although their politicians consult regularly). There is no common newspaper or radio station, but a common television station, ITV Channel Television, as well as a common BBC television news opt-out BBC Channel Islands News.

Geography

The inhabited islands of the Channel Islands are Jersey, Guernsey, Alderncy, Sark, Herm (the main islands); Jethou, Brecqhou (Brechou) and Lihou. All of these except Jersey are in the Bailiwick of Guernsey. In addition there are the following uninhabited islets: the Minquiers, Ecrehous, Les Dirouilles and Les Pierres de Lecq (the Paternosters), are part of the Bailiwick of Jersey, and Burhou and the Casquets lie off Alderney. As a general rule, the larger islands have the-*ey* suffix, and the smaller ones have the-*hou* suffix; this is believed to be from the Old Norse *ey* and *holmr*, respectively.

The Chausey Islands south of Jersey are not generally included in the geographical definition of the Channel Islands but occasionally described as 'French Channel Islands' in English in view of their French jurisdiction. They were historically linked to the Duchy of Normandy, but they are part of the French territory along with continental Normandy, and not part of the British Isles or of the Channel Islands in a political sense.

They are an incorporated part of the commune of Granville (Manche). While popular with visitors from France they are rarely visited by Channel Islanders, as there are no direct transport links from the other islands.

In official Jersey French, the islands are called 'Iles de la Manche', while in France, the term 'Iles anglo-normandes' (Anglo-Norman isles) is used to refer to the British 'Channel Islands' in contrast to other islands in the Channel. Chausey is referred to as an 'Ile normande' (as opposed to *anglo-normande*). 'Iles Normandes' and 'Archipel Normand' have also, historically, been used in Channel Island French to refer to the islands as a whole.

The very large tidal variation provides an environmentally rich inter-tidal zone around the islands, and some sites have received Ramsar Convention designation The waters around the islands include the following:

- The Swinge (between Alderney and Burhou)
- The Little Swinge (between Burhou and Les Nannels)
- La Deroute (between Jersey and Sark, and Jersey and the Cotentin)

- Le Raz Blanchard, or Race of Alderney (between Alderney and the Cotentin)
- The Great Russel (between Sark, Jethou and Herm)
- The Little Russel (between Guernsey, Herm and Jethou)
- Souachehouais (between Le Rigdon and L'Etacq, Jersey)
- Le Gouliot (between Sark and Brecqhou)
- La Percee (between Herm and Jethou).

History

Prehistory

The earliest evidence of human occupation of the Channel Islands has been dated to 25,000 years ago when they were attached to the landmass of continental Europe. The islands became detached by rising sea-levels in the Neolithic period. The numerous dolmens and other archaeological sites extant and recorded in history demonstrate the existence of a population large enough and organised enough to undertake constructions of considerable size and sophistication, such as the burial mound at La Hougue Bie in Jersey or the statue menhirs of Guernsey.

From the Iron Age

Hoards of Armorican coins have been excavated, providing evidence of trade and contact in the Iron Age period. Evidence for Roman settlement is sparse, although evidently the islands were visited by Roman officials and traders. The traditional Latin names of the islands (Caesarea for Jersey; Sarnia for Guernsey, Riduna for Alderney) derive (possibly mistakenly) from the Antonine Itinerary. Gallo-Roman culture was adopted to an unknown extent in the islands. In the 6th century Christian missionaries visited the islands. Samson of Dol, Helier, Marculf and Magloire are among saints associated with the islands. Although originally included within the diocese of Dol, in the 6th century the islands were transferred to the diocese of Coutances, perhaps under the influence of Protextatus. From the beginning of the 9th century Norse raiders appeared on the coasts. Norse settlement succeeded initial attacks, and it is from this period that many place names of Norse origin appear, including the modern names of the islands.

From the Duchy of Normandy

The islands were annexed to the Duchy of Normandy in 933. In 1066, William II of Normandy, a vassal to the king of France, invaded and conquered England, becoming William I of England, also known as William the Conqueror. Since 1204, the loss of the rest of the monarch's lands in mainland Normandy has meant that the Channel Islands have been governed as separate possessions of the Crown.

The islands were invaded by the French in 1338 who held some territory until 1345. Owen of Wales attacked Jersey and Guernsey in 1372, and in 1373 Bertrand du Guesclin besieged Mont Orgueil. Jersey was occupied by the French (as part of the Wars of the Roses) from 1461-1468. In 1483 a Papal Bull decreed that the islands would be neutral during time of war. This privilege of neutrality enabled islanders to trade with both France and England and was respected until 1689 when it was abolished by Order in Council following the Glorious Revolution in Great Britain.

Various attempts to transfer the islands from the diocese of Coutances (to Nantes (1400), Salisbury (1496) and Winchester (1499)) had little effect until an Order in Council of 1569 brought the islands formally into the diocese of Winchester. Control by the bishop of Winchester was ineffectual as the islands had turned overwhelmingly Calvinist and the episcopacy was not restored until 1620 in Jersey and 1663 in Guernsey.

Sark in the 16th century was uninhabited until colonised from Jersey in the 1560s. The grant of sovereigneurship from Elizabeth I of England forms the basis of Sark's constitution today.

Over a dozen windmills are known to have existed in the Channel Isles. They were mostly tower mills used for grinding corn.

From the 17th Century

During the Wars of the Three Kingdoms, Jersey held out strongly for the Royalist cause, providing refuge for Charles, Prince of Wales in 1646 and 1649–1650, while the more strongly Presbyterian Guernsey more generally favoured the parliamentary cause (although Castle Cornet was, on 15 December 1651, the last Royalist stronghold in the British Isles to surrender). The islands acquired commercial and political interests in the North American colonies. Islanders became involved with the Newfoundland fisheries in the 17th century. In recognition for all the help given to him during his exile in Jersey in the 1640s, Charles II gave George Carteret, Bailiff and governor, a large grant of land in the American colonies, which he promptly named New Jersey, now part of the United States of America. Edmund Andros of Guernsey was an early colonial governor in North America, and head of the short-lived Dominion of New England.

20th Century, World War II

The islands were the only part of the British Commonwealth occupied by Germany during the Second World War. The German occupation of 1940–45 was harsh: over 2,000 Islanders were deported by the Germans, Jews sent to concentration camps; partisan resistance and retribution; accusations of collaboration; and slave labour (primarily Russians and eastern Europeans) brought to the islands to build fortifications, with 65,718 landmines laid in Jersey alone.

The British government demilitarised the islands in June 1940 and the Lieutenant-Governors were withdrawn on 21 June, leaving the insular administrations to continue government as best they could under impending military occupation.

Before German troops landed 30 June-4 July 1940, evacuation took place (many young men had already left to join the Allied forces): 6,600 left Jersey (out of 50,000); 17,000 left Guernsey (out of 42,000); the population of Sark remained overwhelmingly; but in Alderney, the entire population, save for six persons, left. In Alderney, the occupying Germans built four concentration camps in which over 700 people died (out of a total inmate population of about 6,000). Due to the destruction of documents, it is impossible to state how many forced workers died in the other islands. These were the only Nazi concentration camps on British soil.

The Royal Navy blockaded the islands from time to time, particularly following the liberation of mainland Normandy in 1944. There was considerable hunger and privation during the five years of German occupation, particularly in the final months when the population was close to starvation. Intense negotiations resulted in some Red Cross humanitarian aid, leading to the arrival of the Red Cross supply ship *Vega* in December 1944.

The end of the occupation only came after VE-Day on 8 May 1945. Jersey and Guernsey were liberated on 9 May 1945. The German garrison in Alderney did not surrender until 16 May 1945 and was one of the last of the Nazi German remnants to surrender. The first evacuees returned on the first sailing from the UK on 23 June, but the population of Alderney was unable to start returning until December 1945.

Post-1945

Following the Liberation of 1945, reconstruction led to a transformation of the economies of the islands, attracting immigration and developing tourism. The legislatures were reformed and non-party governments embarked on social programmes, aided by the incomes from offshore finance which grew rapidly from the 1960s.

The islands decided not to join the European Economic Community when the UK joined, and remain outside.

Since the 1990s declining profitability of agriculture and tourism have challenged the governments of the islands.

Politics

The Channel Islands fall into two separate self-governing bailiwicks. Both the Bailiwick of Guernsey and the Bailiwick of Jersey are British Crown Dependencies, but neither is part of the United Kingdom. They have been part of the Duchy of Normandy since the 10th century and Queen Elizabeth II is often referred to by her traditional and conventional title of Duke of Normandy. However, pursuant to the Treaty of Paris (1259) she is not the Duke in a constitutional capacity and instead governs in her right as Queen. This notwithstanding, it is a matter of local pride for monarchists to treat the situation otherwise: the Loyal Toast at formal dinners is to 'The Queen, our Duke', rather than 'Her Majesty, the Queen' as in the UK.

The Channel Islands are not represented in the UK Parliament but each island has its own primary legislature, known as the States of Guernsey and the States of Jersey, with

Chief Pleas in Sark and the States of Alderney. Laws passed by the States are given Royal Assent by the Queen in Council, to whom the islands' governments are responsible.

The systems of government date from Norman times, which accounts for the names of the legislatures, the States, derived from the Norman 'États' or 'estates' (i.e. the Crown, the Church, and the people). The States have evolved over the centuries into democratic parliaments.

A bailiwick is a territory administered by a bailiff. The Bailiff in each bailiwick is the civil head, presiding officer of the States, and also head of the judiciary.

In 2001, responsibility for links between the Channel Islands (together with the Isle of Man) and the Crown passed from the Home Secretary to the Lord Chancellor's Department, replaced in 2003 by the Department of Constitutional Affairs and in 2007 by the Ministry of Justice.

In addition, Acts of the UK Parliament may be extended to any of the Channel Islands by Order-in-Council (thus giving the UK Government some responsibility for good governance in the islands). By constitutional convention this is only done at the request of the Insular Authorities, and has become a rare option (thus giving the Insular Authorities themselves the responsibility for good governance in the islands), the islands usually preferring nowadays to pass localised versions of laws giving effect to international treaties.

Matters reserved to the Crown (i.e. acting through the United Kingdom Government) are limited to defence, citizenship, and diplomatic representation. The islands are not bound by treaties concluded by the United Kingdom (unless they so request) and may separately conclude treaties with foreign governments (except concerning matters reserved to the Crown). The United Kingdom conceded at the end of the 20th century that the islands may establish direct political (non-diplomatic) contacts with foreign governments to avoid the situation whereby British embassies were obliged to pass on communications from the governments of the Bailiwicks that were in conflict with United Kingdom government policy.

The islands are not part of the European Union, but are part of the Customs Territory of the European Community, by

Islanders are full British citizens, and therefore European Citizens. Any British citizen who applies for a passport in Jersey or Guernsey receives a passport bearing the words "British Islands, Bailiwick of Jersey" or "British Islands, Bailiwick of Guernsey". Under the provisions of Protocol Three, Channel Islanders who do not have a close connection with the UK (no parent or grandparent from the UK, and have never been resident in Great Britain or Northern Ireland for any five-year period) do not automatically benefit from the EU provisions on free movement within the EU and consequently their passports receive an endorsement to that effect. This only affects a minority of islanders.

Under the Interpretation Act 1978, the Channel Islands are deemed to be part of the British Islands, not to be confused with the British Isles. But according to the British Nationality Act 1981 the United Kingdom means Great Britain, Northern Ireland and the Channel Islands and the Isle of Man, taken together.

Both Bailiwicks are members of the British-Irish Council, and Jerriais and guernesiais are recognised regional languages of the Isles.

The legal courts are separate; separate courts of appeal have been in place since 1961. Among the legal heritage from Norman law is the Clameur de Haro.

Economy

Tourism is the major industry in the smaller islands (with some agriculture). Jersey and Guernsey have, since the 1960s, relied on financial services. Guernsey's horticultural and greenhouse activities have been more significant than in Jersey, and Guernsey has maintained light industry as a higher proportion of its economy than Jersey. Jersey's economy since the 1980s has been substantially more reliant on finance.

Both Bailiwicks issue their own banknotes and coins, which circulate freely in all the islands alongside UK coinage and Bank of England and Scottish banknotes.

There are many exports, largely consisting of crafted goods and farmed produce. The Genuine Jersey Products Association certify products as being locally made/sourced.

Transport and Communications

Since 1969, Jersey and Guernsey have operated postal administrations independently of the UK's Royal Mail, with their own postage stamps, which can only be used for postage in their respective Bailiwicks. UK stamps are no longer valid, but mail to the islands, and to the Isle of Man, is still charged at UK inland rates. However, it was not until the early 1990s that the islands joined the UK's postcode system, Jersey postcodes using the initials JE and Guernsey using GY.

Alderney, Guernsey and Jersey are connected by what could be called an aerial bus service operated and competed on by Blue Islands and Aurigny. The latter use the incredibly rare Britten-Norman Trislander on the route while the former use Britten-Norman Islanders and BAE Jetstream 32s.

The islands are connected to the radio and television system of the UK. They are part of BBC Channel Islands, and have since 2000 had regular opt-outs from the main Spotlight programme: 15 minutes at 18.30 and a full late bulletin at 22.25. There are also two local BBC radio stations, BBC Radio Guernsey and BBC Radio Jersey.

Alderney has its own radio station, QUAY-FM, which operates in the summer tourist season and at Christmas.

The islands have had their own ITV franchise, Channel Television, since September 1962. The islands will switch to digital-only transmissions in November 2010.

Jersey always operated its own telephone services independently of the UK's national systems, but Guernsey did not establish its own telephone services until 1969. Both islands still form part of the UK telephone numbering plan, but Ofcom in the UK does not have responsibility for telecommunications regulatory and licensing issues on the islands. (It is responsible for wireless telegraphy licensing throughout the islands, and by agreement, for broadcasting regulation in the two large islands only).

The Channel Islands have their own country-code top-level-domains (ccTLDs) on the Internet, managed by Channelisles.net. The ccTLDs are.GG for the Bailiwick of Guernsey (including Alderney and Sark) and JE for the Bailiwick of Jersey. The country codes first appeared on the Internet in

1996 after Jon Postel agreed with Nigel Roberts of Island Networks to add four codes (GG, JE as well as IM and AC) to the IANA list of TLDs. The codes for the Channel Islands and for the Isle of Man were entered on to the official United Nations ISO-3166 list in 2006.

Alderney has a large and growing internet gambling industry with a full regulatory authority in operation.

Each of the three largest islands has a distinct vehicle registration scheme:

- Guernsey (GBG): simply a number, up to five digits;
- Jersey (GBJ): J followed by up to six digits (JSY vanity plates are also issued);
- Alderney (GBA): *AY* followed by up to five digits (four digits are the most that have been used, as redundant numbers are re-issued).

In Sark, where most motor traffic is prohibited, the few vehicles on the island – nearly all tractors – do not display plates. Bicycles display cardboard tax discs.

In the 1960s, names used for the cross-Channel ferries plying the mail route between the islands and Weymouth, Dorset were taken from the popular Latin names for the islands: "Caesarea" (Jersey), "Sarnia" (Guernsey) and "Riduna" (Alderney).

Today, the ferry route between the Channel Islands and the UK is operated by Condor Ferries from both St. Helier, Jersey and St. Peter Port, Guernsey using high speed catamarans fast craft to Weymouth, and Poole in the UK. A regular passenger ferry service on the Commodore Clipper goes from both Channel Island ports to Portsmouth daily which carries both passengers and freight.

Ferry services to mainland Normandy are operated by Manche Iles Express, and a service between Jersey and Saint Malo is also operated by Compagnie Corsaire. Condor also operate a service to Saint Malo.

Culture

Culturally, the Norman language predominated in the islands until the 19th century, when increasing influence from English-speaking settlers and easier transport links led to

Anglicisation. There are four main dialects/languages of Norman in the islands, Auregnais (Alderney, extinct in late 20th century), Dgernesiais (Guernsey), Jerriais (Jersey) and Sercquiais (Sark, an offshoot of Jerriais).

Victor Hugo spent many years in exile, first in Jersey and then in Guernsey, where he wrote *Les Miserables*. Guernsey is also the setting of Hugo's later novel, *Les Travailleurs De La Mer* (*The Toilers of the Sea*). A "Guernsey-man" also makes an appearance in Herman Melville's *Moby-Dick*.

The annual "Muratti", the inter-island football match, is considered the sporting event of the year—although, due to broadcast coverage, it no longer attracts the crowds of spectators, travelling between the islands, that occurred during the 20th century.

Channel Island sportsmen and women compete in the Commonwealth Games for their respective islands and the islands have also been enthusiastic supporters of the Island Games. Shooting is a popular sport, in which islanders have won Commonwealth medals.

Guernsey's traditional colour for sporting and other purposes is green and Jersey's is red.

The main islanders have traditional animal nicknames:

- Guernsey: *les anes* ("donkeys" in French and Norman): the steepness of St. Peter Port streets required beasts of burden, but Guernsey people also claim it is a symbol of their strength of character – which Jersey people traditionally interpret as stubbornness.
- Jersey: *les crapauds* ("toads" in French and Jerriais): Jersey has toads and snakes that Guernsey lacks.
- Sark: *les corbins* ("crows" in Sercquiais, Dgernesiais and Jerriais, *les corbeaux* in French): crows could be seen from sea on the island's coast.
- Alderney: *les lapins* ("rabbits" in French and Auregnais): the island is noted for its warrens.

Christianity was brought to the islands around the 6th century; according to tradition, Jersey was evangelised by Saint Helier, Guernsey by Saint Samson of Dol and other smaller islands were occupied at various times by monastic communities representing strands of Celtic Christianity. At the Reformation, the islands turned Calvinist under the influence of an influx of French-language pamphlets published in Geneva. Anglicanism was imposed in the 17th century, but the Non-Conformist tendency re-emerged with a strong adoption of Methodism. The presence of long-term Catholic communities from France and seasonal workers from Brittany and Normandy added to the mix of denominations among the population.

Other Islands in the English Channel

There are other islands in other stretches of the English Channel that are not traditionally included within the grouping of Channel Islands. Among these are Ouessant/Ushant, Brehat, Ile de Batz, Chausey, Grande-Ile, Tatihou and Iles Saint-Marcouf (under French jurisdiction) and the Isle of Wight and Isles of Scilly (both part of England and under UK jurisdiction).

4

List of Islands in the British Isles

Geology

The British Isles lie at the juncture of several regions with past episodes of tectonic mountain building. These orogenic belts form a complex geology which records a huge and varied span of earth history. Of particular note was the Caledonian Orogeny during the Ordovician Period, ca. 488–444 Ma and early Silurian period, when the craton Baltica collided with the terrane Avalonia to form the mountains and hills in northern Britain and Ireland. Baltica formed roughly the north western half of Ireland and Scotland. Further collisions caused the Variscan orogeny in the Devonian and Carboniferous periods, forming the hills of Munster, south-west England, and south Wales. Over the last 500 million years the land which forms the islands has drifted northwest from around 30°S, crossing the equator around 370 million years ago to reach its present northern latitude.

The islands have been shaped by numerous glaciations during the Quaternary Period, the most recent being the Devensian. As this ended, the central Irish Sea was de-glaciated (whether or not there was a land bridge between Great Britain and Ireland at this time is somewhat disputed, though there was certainly a single ice sheet covering the entire sea) and the English Channel flooded, with sea levels rising to current levels some 4,000 to 5,000 years ago, leaving the British Isles in their current form.

The islands' geology is highly complex, though there are large numbers of limestone and chalk rocks that formed in the Permian and Triassic periods. The west coasts of Ireland and northern Great Britain that directly face the Atlantic Ocean are generally characterized by long peninsulas, and headlands and bays; the internal and eastern coasts are "smoother".

Geology of Great Britain

Great Britain shows a rich variety of landscape across the constituent countries of England, Wales and Scotland. Landforms and rocks of all geological ages are represented, from Precambrian onwards.

Seismographical Results

Seismographical research shows that the crust of the Earth below Great Britain is between 27 and 35 km (17 to 22 miles) thick. The oldest rocks are found at the surface in north west Scotland and are more than half as old as the planet. They are thought to underlie much of Great Britain and Ireland (although boreholes have only penetrated the first few kilometres), but next appear extensively at the surface in Brittany and the Channel Islands. The youngest rocks are found in south east England.

Bedrock

The bedrock consists of layers formed over vast periods of time. These were laid down as the climate changed, the landmasses moved due to plate tectonics, and the land and sea levels rose or fell. From time to time horizontal forces caused the rock to undergo deformation, folding the layers of rock to form mountains which have since been eroded and overlain with other layers. To further complicate the geology, the land has also been subject to periods of earthquakes and volcanic activity.

Deposits by Glaciers

Overprinted on this bedrock geology ("solid geology" in the terminology of maps) is a distribution of soils and fragmental material deposited by glaciers (boulder clay, and other forms of glacial drift in the recent past. "Drift" geology is often more important than "solid" geology when considering such as building works, drainage, siting water boreholes, sand and gravel resources and soil fertility. Although "drift" strictly refers to glacial and fluvio-glacial deposits, the term on geological maps has traditionally included other material including alluvium, river terraces, etc. Recent maps use the terms "bedrock" and "superficial" in place of "solid" and "drift".

Geological History

Proterozoic Era

The Gneisses, the oldest rocks in Great Britain or Ireland, date from at least 2,700 Ma (Ma = million years ago) in the Archaean period of this era, the Earth itself being about 4,600 Ma old. They are found in the far north west of Scotland and in the Hebrides, with a few small outcrops elsewhere. Formed from rock originally deposited at the surface of the planet, the rocks were later buried deep in the Earth's crust and metamorphosed into crystalline gneiss.

South of the Gneisses are a complex mixture of rocks forming the North West Highlands and Grampian Highlands in Scotland, as well as the Connemara, Donegal and Mayo mountains of Ireland. These are essentially the remains of folded sedimentary rocks that were originally 25 km thick, deposited over the gneiss on what was then the floor of the Iapetus Ocean. The process started in about 1,000 Ma, with a notable 7 km thick layer of Torridon Sandstone being deposited about 800 Ma, as well as the debris deposited by an ice sheet 670 Ma.

Palaeomagnetic evidence indicates that 520 Ma, what is now the UK was split between two continents, separated by 7000 km (4500 miles) of ocean. The north of Scotland was located at about 20° south of the equator on the continent of Laurentia near the Tropic of Capricorn, while the rest of the country was at about 60° south on the continent of Gondwana near the Antarctic Circle.

In Gondwana, England and Wales were near a subduction zone. Both countries were largely submerged under a shallow sea studded with volcanic islands. The remains of these islands underlie much of central England with small outcrops visible in many places. Around 600 Ma, the Cadomian Orogeny (mountain building period) caused the English and Welsh landscape to be transformed into a mountainous region, along with much of north west Europe.

Palaeozoic Era

Cambrian Period

In the early Cambrian period the volcanoes and mountains of England and Wales were eroded as the land became flooded by a rise in sea level, and new layers of sediment were laid down. Much of central England formed a stable block of crust which has remained largely undeformed ever since. Sandstones were deposited in the north of Scotland. The first animals with hard shells evolved at this time, consequently, fossils become much more common in rocks formed during this and later periods.

Ordovician Period

500 million years ago, in the Ordovician period, southern Britain, the east coast of North America and south-east Newfoundland broke away from Gondwana to form the continent of Avalonia, which by 440 Ma had drifted (by the mechanisms of plate tectonics) to about 30° south.

During this period north Wales and the south Mayo area of Ireland were subject to volcanic activity. The remains of these volcanoes are still visible, one example of which is Rhobell Fawr dating from 510 Ma. Large quantities of volcanic lava and ash known as the Borrowdale Volcanics covered both Wales and the Lake District and this can still be seen in the form of mountains such as Helvellyn and Scafell Pike.

The Ordovician also saw the formation of the Welsh Skiddaw slate deposits around 500 Ma.

Silurian Period

Deposition continued into the early part of the Silurian period, with mudstones and sandstones being laid down, notably in Wales.

Avalonia had now joined with the continent of Baltica, and the combined landmass collided with Laurentia at about 20° south between 425 and 400 Ma, joining the southern and northern halves of Great Britain together. The resulting Caledonian Orogeny produced an Alpine-style mountain range in much of north and west Britain. The continental collision was probably at an oblique angle rather than a head-on collision, and this probably led to movement along strike-slip faults trending north-east to south-west across Scotland, the Great Glen Fault being the best example (some of these fault zones may have been old lines of weakness from earlier earth movements).

Volcanic ashes and lavas deposited during the Silurian are still found in the Mendip Hills and in Pembrokeshire.

Devonian Period

The collision between continents continued during the Devonian period, with continuing uplift, and more volcanic deposits such as those now forming Ben Nevis. Sea levels varied considerably, with the coastline advancing and retreating from north to south across England. The uplifted region was gradually eroded down, resulting in the deposition of numerous sedimentary rock layers in lowlands and seas. The rocks of marine origin in the county of Devon gave the period its name, though deposits of this age are found in many other places, such as the Brecon Beacons, the Midland Valley of Scotland, and the Orkney Islands. Most of these are of terrestrial origin and are informally known as the Old Red Sandstone.

The Caledonian mountains had largely been eroded away by the end of the period during which the country would have experienced an arid desert climate as it was located close to the equator at between 10° and 15° south.

Carboniferous Period

Around 360 Ma during the Carboniferous period Great Britain was lying at the equator, covered by the warm shallow waters of the Rheic Ocean, during which time the Carboniferous Limestone was deposited, as found in the Mendip Hills, north and south Wales, in the Peak District of Derbyshire, north Lancashire, the northern Pennines and southeast Scotland. Caves developed more recently in the limestone of some of these areas by the action of carbonic acid and other organic acids in rainwater and groundwater.

These were followed by dark marine shales, siltstones, and coarse sandstones of the Millstone Grit. Later, river deltas formed and the sediments deposited were colonised by swamps and rain forest. It was in this environment that the cyclic Coal Measures were formed, the source of the majority of Britain's extensive coal reserves that powered the Industrial Revolution. Coal can be found in many areas of Britain and Ireland, as far north as the Midland Valley of Scotland, as far south as Kent and in Ireland, although coal mining has largely been concentrated in the English midlands, northern England and Wales.

Throughout the period, southwest England in particular was affected by the collision of continental plates. The mountain building period known as the Variscan orogeny, which occurred around 280 Ma, caused major deformation in south west England.

Towards the end of this period granite was formed beneath the overlying rocks of Devon and Cornwall, now exposed at Dartmoor and Bodmin Moor, giving rise to mineralised deposits of copper and tin. The general region of Variscan folding was south of an east–west line roughly from south Pembrokeshire to Kent. The main tectonic pressure was from the south or south-east, and may have resulted in dextral strike-slip faulting. The Devon-Cornwall massif may originally have been some distance further east, then to be moved westwards. Lesser Variscan folding took place as far north as Derbyshire and Berwick-upon-Tweed.

By the end of the Carboniferous period the various continents of the Earth had fused to form the super-continent of Pangaea. Britain was located in the interior of Pangea where it was again subject to a hot arid desert climate with frequent flash floods leaving deposits that formed beds of red sedimentary rock, somewhat similar to the later, Triassic New Red Sandstone.

Permian Period

The Permian was characterised for 30 million years by arid desert conditions and the erosion of the land that had uplifted in the Variscan Orogeny, southwest England and adjacent areas of the present-day English Channel. Later, much of Great Britain was submerged in shallow waters as the polar ice sheets melted and the Tethys Ocean and Zechstein Sea formed, depositing shale, limestone, gravel, and marl, before finally receding to leave a flat desert with salt pans.

Mesozoic Era

Triassic Period

As Pangaea drifted during the Triassic, Great Britain moved away from the equator until it was between 20° and 30° north. Red beds, including sandstones and red mudstones form the main sediments of the New Red Sandstone. The remnants of the Variscan uplands in France to the south were eroded down, resulting in layers of the New Red Sandstone being deposited across central England, and in faulted basins in Cheshire and the Irish Sea. A basin developed in the Hampshire region around this time. Rifting occurred within and around Britain and Ireland, prior to the breakup of the super-continent in the Jurassic period.

Rock fragments found near Bristol appear to indicate that in 214 Ma Great Britain was showered with a fine layer of debris from an asteroid impact at the Manicouagan Impact Crater in Canada, although this is still being debated.

Jurassic Period

As the Jurassic period began, Pangaea began to break up, sea levels rose and Britain and Ireland drifted on the Eurasian Plate to between 30° and 40° north. With much of the British Isles under water again, sedimentary rocks were deposited and can now be found underlying much of England from the Cleveland Hills of Yorkshire to the Jurassic Coast in Dorset. These include sandstones, greensands, oolitic limestone of the Cotswold Hills, corallian limestone of the Vale of White Horse and the Isle of Portland.

The burial of algae and bacteria below the mud of the sea floor during this time resulted in the formation of North Sea oil and natural gas, much of it trapped in overlying sandstone by salt deposits formed as the sea levels fell to form the swamps and salty lakes and lagoons that were home to dinosaurs.

Cretaceous Period

The modern continents having formed, the Cretaceous saw the formation of the Atlantic Ocean, gradually separating northern Scotland from North America. The land underwent a series of uplifts to form a fertile plain. After 20 million years or so, the seas started to flood the land again until much of

Britain and Ireland were again below the sea, though sea levels frequently changed. Chalk and flints were deposited over much of Great Britain, now notably exposed at the White Cliffs of Dover and the Seven Sisters, and also forming Salisbury Plain. The high sea levels left only small areas of land exposed, which accounts for the general lack of land-origin sand, mud or clay sediments found from around this time. Some of the late Cretaceous strata are, in fact, almost pure chalk.

Cenozoic Era

Palaeogene Period

In the early Palaeogene period between 63 and 52 Ma, the last volcanic rocks in Great Britain were formed. The major eruptions at this time produced the Antrim Plateau, the basaltic columns of the Giant's Causeway and Lundy Island in the Bristol Channel. The Alpine Orogeny that took place about 50 Ma was responsible for the shaping of the London Basin syncline, the Weald-Artois Anticline to the south, the North Downs, South Downs and Chiltern Hills. During the period the North Sea formed, Britain was uplifted. Some of this uplift was along old lines of weakness left from the Caledonian and Variscan Orogenies long before. The uplifted areas were then eroded, and further sediments, such as the London Clay, were deposited over southern England, while the English Channel was characterised by mud flats and river-deposited sands. Much of the midlands and north of England may have been covered by Jurassic and Cretaceous deposits at the start of the Palaeogene, but these were lost through erosion. By 35 Ma the landscape was colonised by trees such as beech, oak, redwood and palm, along with grasses.

Neogene Period

In the Miocene and Pliocene epochs of the Neogene, further uplift and erosion occurred, particularly in Wales, the Pennines, and the Scottish Highlands. Plant and animal types developed into their modern forms, and by about 2 million years ago the landscape would have been broadly recognisable today.

Pleistocene Epoch

The major changes during the Pleistocene were brought about by several recent ice ages. The most severe was the

Anglian Stage, with ice up to 1,000 m (3300 ft) thick that reached as far south as London and Bristol. This took place between about 478,000 to 424,000 years ago, and was responsible for the diversion of the River Thames onto its present course.

There is extensive evidence in the form of stone tools that southern England was colonised by human populations during the warm Hoxnian Stage that followed the Anglian Glaciation. It is possible that the English Channel repeatedly opened and closed during this period, causing Britain to become an island from time to time. The oldest human fossils found to date in British Isles, which include the skull of Swanscombe Man from 250,000 years ago, and the earlier Clactonian Man, also date from this period.

The Wolstonian Stage, between about 352,000 to 130,000 years ago, which is thought to have peaked around 150,000 years ago, was named after the town of Wolston south of Birmingham which is thought to mark the southern limit of the ice.

The Wolstonian Stage was followed by the Ipswichian Stage, during which hippopotamus are known to have lived as far north as Leeds.

During the most recent Devensian glaciation, which is thought to have started around 115,000 years ago, peaked around 20,000 years ago and ended a mere 10,000 years ago, the Usk valley and Wye valley were eroded by glaciers, with the ice sheet itself reaching south to Birmingham. The oldest human remains in Britain or Ireland, the Red Lady of Paviland (29,000 years old) date from this time. It is thought that the country was eventually abandoned as the ice sheet reached its peak, being recolonised as it retreated. By 5,000 years ago it is thought that Great Britain was warmer than it is at present.

Among the features left behind by the ice are the fjords of the west coast of Scotland, the U shaped valleys of the Lake District and erratics (blocks of rock) that have been transported from the Oslo region of Norway and deposited on the coast of Yorkshire.

Holocene Epoch

The last twelve thousand years are known as the Holocene Epoch. The most significant new geological features during this period have been the deposition of peat in Ireland and Scotland, and in coastal areas of England. Many of the lowland deposits, such as the Somerset Levels, The Fens and Romney Marsh have recently been artificially drained

Since humans began clearing the forest during the new stone age, most of the land has now been deforested, speeding the natural processes of erosion. Large quantities of stone, gravel and clay are extracted each year, and by 2000 11% of England was covered by roads or buildings.

At the present time Scotland is continuing to rise as a result of the weight of Devensian ice being lifted. The rest of Britain is sinking, generally estimated at 1 mm (1/25 inch) per year, with the London area sinking at double the speed partly due to the continuing compaction of the recent clay deposits.

In addition, rises in sea level thought to be due to global warming appear likely to make low lying areas of land increasingly susceptible to flooding, while in some areas the coastline continues to erode at a geologically rapid rate.

Great Britain continues to be subject to several very minor earthquakes each month, and more occasional light to moderate ones. During the 20th century 25 earthquakes with a moment magnitude of 4.5 to 6.1 were felt, many of them originating within the Isles themselves.

Geological Features

- Cheddar Gorge-the largest gorge in Great Britain
- Jurassic Coast-a UNESCO World Heritage Site
- Great Glen Fault
- Highland Boundary Fault
- Southern Uplands Fault
- Midland Valley of Scotland
- Fingal's Cave-columnar Tertiary basaltic formation (similar to the Giant's Causeway in Northern Ireland)
- Salisbury Plain
- Tees-Exe line
- Whin Sill
- Siccar Point in Berwickshire provided early proof of the immense age of the Earth

- The Moine Thrust was the first thrust belt to be discovered by geologists
- London Basin
- Hampshire Basin
- London-Brabant Massif
- Dartmoor-one of a series of Moors in the South West of England developing Tors on a granitic batholith
- Lizard Complex-an ancient piece of oceanic crust, onshore (i.e. an Ophiolite).

Geological Resources

- Coal
- North Sea oil
- Sand and Gravel
- China Clay
- Copper and Tin
- Stone
- Geothermal energy
- Onshore oil.

Events

- Storegga Slides, caused a tsunami circa 6100 BC
- Bristol Channel floods, 1607 was caused by a tsunami
- 1755 Lisbon earthquake caused a tsunami that hit Cornwall
- Colchester earthquake, 1884
- 2002 Dudley earthquake
- 2008 Lincolnshire earthquake.

Transport

Heathrow is Europe's busiest airport in terms of passenger traffic and the Dublin-London route is the busiest air route in Europe. The English Channel and the southern North Sea are the busiest seaways in the world. The Channel Tunnel, opened in 1994, links Great Britain to France and is the second-longest rail tunnel in the world. The idea of building a tunnel under the Irish Sea has been raised since 1895, when it was first investigated. Several potential Irish Sea tunnel projects have been proposed, most recently the *Tusker Tunnel* between the ports of Rosslare and Fishguard proposed by The Institute of Engineers of Ireland in 2004. A rail tunnel was proposed in 1997 on a different route, between Dublin and Holyhead, by British engineering firm Symonds. Either tunnel, at 80 km (50 mi), would be by far the longest in the world, and would cost an estimated €20 billion. A proposal in 2007, estimated the cost of building a bridge from County Antrim in Northern Ireland to Galloway in Scotland at £3.5bn (€5bn).

Demographics

The demographics of the British Isles shows a generally high density of population in England, which accounts for almost 80% of the total population of the islands. In Ireland, Northern Ireland, Scotland and Wales high density of population is limited to areas around, or close to, their respective capitals. Major population centres (greater than one million people) exist in the following areas:

- Greater London Urban Area (8.5 million)
    - London metropolitan area (12—14 million)
- Greater Manchester Urban Area (2.5 million)
- West Midlands conurbation (2.3 million)
- West Yorkshire Urban Area (2.1 million)
- Hampshire (1.7 million)
- Greater Glasgow (1.7 million)
- Greater Dublin Area (1.6 million)
- Lancashire (1.4 million)
- South Yorkshire (1.2 million)
- Tyne and Wear (1.1 million).

The population of England has risen steadily throughout its history, while the populations of Scotland and Wales have shown little increase during the twentieth century-the population of Scotland remaining unchanged since 1951. Ireland, which for most of its history comprised a population proportionate to its land area (one third of the total population) has, since the Great Famine, fallen to less than one tenth of the population of the British Isles. The famine, which caused a century-long population decline, drastically reduced the Irish population and permanently altered the demographic make-up of the British Isles. On a global scale this disaster led to the creation of an Irish diaspora that numbers fifteen times the current population of the island

Political Co-operation within the Islands

Between 1801 and 1922, Great Britain and Ireland together formed the United Kingdom. In 1922, twenty-six counties of Ireland seceded from the United Kingdom following the Irish War of Independence and the Anglo-Irish Treaty; the remaining six counties, mainly in the northeast of the island, became known as Northern Ireland under the Government of Ireland Act, 1920. Both states, but not the Isle of Man or the Channel Islands, are members of the European Union.

However, political cooperation between the United Kingdom and the Republic of Ireland exists on some levels:

- Travel. Since Irish partition an informal free-travel area has continued to exist across the entire region; in 1997 it was formally recognised by the European Union, in the Amsterdam Treaty, as the Common Travel Area.
- Voting rights. No part of the British Isles considers a citizen of any other part as an 'alien'. This pre-dates and goes much further than that required by European Union law, and gives common voting rights to all citizens of the jurisdictions within the archipelago. Exceptions to this are presidential elections and referendums in the Republic of Ireland, for which there is no comparable franchise in the other states. Other EU nationals may only vote in local and European Parliament elections while resident in either the UK or Ireland. A 2008 UK Ministry of Justice report proposed to end this arrangement arguing that, "the right to vote is one of the hallmarks of the political status of citizens; it is not a means of expressing closeness between countries."
- Diplomatic. Bilateral agreements allow UK embassies to act as an Irish consulate when Ireland is not represented in a particular country.
- Northern Ireland. Citizens of Northern Ireland are entitled to the choice of Irish or British citizenship or both.

- The British-Irish Council was set up in 1999 following the 1998 Belfast Agreement. This body is made up of all political entities across the islands, both the sovereign governments of Ireland and the United Kingdom, the devolved governments of Northern Ireland, Scotland and Wales, and the dependencies of Guernsey, Jersey and the Isle of Man. It has no executive authority but meets biannually to discuss issues of mutual importance, currently restricted to the misuse of drugs, the environment, the knowledge economy, social inclusion, tele-medicine, tourism, transport and national languages of the participants. During the February 2008 meeting of the Council, it was agreed to set-up a standing secretariat that would serve as a permanent 'civil service' for the Council.
- The British-Irish Inter-Parliamentary Body was established in 1990. Originally it comprised 25 members of the Oireachtas, the Irish parliament, and 25 members of the parliament of the United Kingdom, with the purpose of building mutual understanding between members of both legislature. Since then the role and scope of the body has been expanded with the addition of five representatives from the Scottish Parliament, five from the National Assembly for Wales and five from the Northern Ireland Assembly. One member is also taken from the States of Jersey, one from the States of Guernsey and one from the High Court of Tynwald (Isle of Man). With no executive powers, it may investigate and collect witness evidence from the public on matters of mutual concern to its members, these have in the past ranged from issues such as the delivery of health services to rural populations, to the Sellafield nuclear facility, to the mutual recognition of penalty points against drivers across the British Isles. Reports on its findings are presented to the governments of Ireland and the United Kingdom. Leading on from developments in the British-Irish Council, the chair of the Body, Niall Blaney, has suggested a name-change and that the body should shadow the British-Irish Council's work.

Languages

The linguistic heritage of the British Isles is rich, with twelve languages from six groups across four branches of the Indo-European family. The Insular Celtic languages of the Goidelic sub-group (Irish, Manx and Scottish Gaelic) and the Brythonic sub-group (Cornish, Welsh and Breton, spoken in north-western France) are the only remaining Celtic languages-the last of their continental relations becoming extinct before the 7th century. The Norman languages of Guernesiais, Jerriais and Sarkese are spoken in the Channel Islands are similar to French. A cant, called Shelta, is a language spoken by Irish Travellers, often as a means to conceal meaning from those outside the group. However, English, sometimes in the form of Scots, is the dominant language, with few monoglots remaining in the other languages of the region. The Norn language of Orkney and Shetland became extinct around 1880.

Until perhaps 1950 the use of languages other than English roughly coincided with the major ethno-cultural regions in the British Isles. As such, many of them, especially the Celtic languages, became intertwined with national movements in these areas, seeking either greater independence from the parliament of the United Kingdom, seated in England, or complete secession. The common history of these languages was one of sharp decline in the mid-19th century, prompted by centuries of economic deprivation and official policy to discourage their use in favour of English. However, since the mid-twentieth century there has been somewhat of a revival of interest in maintaining and using them. Celtic-language medium schools are available throughout Ireland, Scotland and Wales to such an extent that it is now possible to receive all formal education, up to and including third-level education, through a Celtic language. Instruction in Irish and Welsh is compulsory in all schools in the Republic of Ireland and Wales respectively. In the Isle of Man, Manx in taught in all schools, although it is not compulsory, and there is one Manx-medium school. The respective languages are official languages of state in Ireland, the Isle of Man, Scotland and Wales, with equal status with respect to English. In the Channel Islands French is a legislative and administrative language. Since 2007, Irish is a working language of the European Union.

During the last 60 years there has been a great deal of immigration into Great Britain (less into Ireland). As a result a number of languages not formerly found in the British Isles are in regular use. Polish, Punjabi, and Hindustani (inc Urdu & Hindi), are each probably the first language of over 1 million residents, and a number of other languages are regularly spoken by substantial numbers of persons. Even in provincial areas it has become common for local government to publish information to residents in ten or so languages, and in the largest city, London, the first language of about 20% of the population is neither English nor an indigenous Celtic language. Cornish and the Norman languages of Guernesiais, Jerriais and Sarkese are far less supported. In Jersey, a language office (*L'Office du Jerriais*) is funded to provide education services for Jerriais in schools and other language services, while in Guernsey there is a language officer and Guernesiais is taught in some schools on a volunteer basis. Of the four, only Cornish is recognised officially under the European Charter for Regional or Minority Languages, and it is taught in some schools as an optional modern language. Guernesiais and Jerriais are recognised as regional languages by the British and Irish governments within the framework of the British-Irish Council. Scots, as either a dialect of or a closely related language to English, is similarly recognised by the European Charter, the British-Irish Council, and as "part of the cultural wealth of the island of Ireland" under the Good Friday Agreement. However, it is without official status as a language of state in Scotland, where English is used in its place.

Shelta, spoken by the ethnic minority Irish Travellers, is thought to be spoken by 6,000–25,000 people, according to varying sources. Although evidence suggests that it existed as far back as the 13th century, as a secret language, it was only discovered at the end of the 19th century. It is without any official status, despite being thought to have 86,000 speakers worldwide, mostly in the USA.

Sport

A number of sports are popular throughout the British Isles, the most prominent of which is association football. While this is organised separately in different national associations, leagues and national teams, even within the UK, it is a common passion in all parts of the islands. Rugby union is also widely enjoyed across the islands. The British and Irish Lions is a team made up of players from England, Ireland, Scotland and Wales that undertakes tours of the southern hemisphere rugby playing nations every four years. This team was formerly known as the *British Isles* and the *British Lions*, but has been called the *British and Irish Lions* since 2001. Ireland play as a united team, represented by players from both Northern Ireland and the Republic. The four national rugby teams from Great Britain and Ireland play each other each year for the Triple Crown as part of the Six Nations Championship. Also since 2001 the professional club teams of Ireland, Scotland and Wales have competed together in the Magners League.

The Ryder Cup in golf was played between a United States team and a team representing Great Britain and Ireland. From 1979 onwards this was expanded to include the whole of Europe.

Popular Culture

The United Kingdom and Ireland have separate media, although British television, newspapers and magazines are widely available in Ireland, giving people in Ireland a high level of familiarity with cultural matters in Great Britain.

A few cultural events are organised for the island group as a whole. For example, the Costa Book Awards are awarded to authors resident in the UK or Ireland. The Man Booker Prize is awarded to authors from the Commonwealth of Nations or Ireland. The Mercury Music Prize is handed out every year to the best album from a British or Irish musician or group.

Many other bodies are organised throughout the islands as a whole; for example the Samaritans which is deliberately organised without regard to national boundaries on the basis that a service which is not political or religious should not recognise sectarian or political divisions. The RNLI is also organised throughout the islands as a whole, covering both the United Kingdom and Ireland.

5

France

France, officially the French Republic, is a state in Western Europe with several of its overseas territories and islands located on other continents and in the Indian, Pacific, and Atlantic oceans. Metropolitan France extends from the Mediterranean Sea to the English Channel and the North Sea, and from the Rhine to the Atlantic Ocean. It is often referred to as *L'Hexagone* ("The Hexagon") because of the geometric shape of its territory. It is bordered (clockwise from the north) by Belgium, Luxembourg, Germany, Switzerland, Italy, Monaco, Spain and Andorra. France's overseas departments and collectivities also share land borders with Brazil and Suriname (bordering French Guiana), and the Netherlands Antilles (bordering Saint-Martin). France is linked to the United Kingdom by the Channel Tunnel, which passes underneath the English Channel.

France is a member state of the European Union, the largest one by area. It is also the third largest in Europe behind Russia and Ukraine. It would be second if its extra-European territories like French Guiana were included. France has been a major power for several centuries with strong economic, cultural, military and political influence. During the 17th and 18th centuries, France colonised great parts of North America; during the 19th and early 20th centuries, France built the second largest empire of the time, including large portions of North, West and Central Africa, Southeast Asia, and many Pacific islands. France is a unitary semi-presidential republic with its main ideals expressed in the Declaration of the Rights of Man and of the Citizen. France is one of the most developed countries and possesses the fifth largest economy by nominal GDP and seventh largest economy by purchasing power parity. France enjoys a high standard of living as well as a high public education level, and has also one of the world's highest life expectancies, It is the most visited country in the world, receiving 82 million foreign tourists annually. France is one of the founding members of the European Union. It is also a founding member of the United Nations, and a member of the Francophonie, the G8, G20, NATO, OECD, WTO, and the Latin Union. It is one of the five permanent members of the UN Security Council and possesses the third largest nuclear weapons stockpile in the world.

Origin of the Name France

The name "France" comes from Latin *Francia*, which literally means "land of the Franks," or "Frankland". There are various theories as to the origin of the name of the Franks. One is that it is derived from the Proto-Germanic word *frankon* which translates as *javelin* or *lance* as the throwing axe of the Franks was known as a francisca.

Another proposed etymology is that in an ancient Germanic language, Frank means *free* as opposed to slave. This usage still survives in the name of the national currency prior to the adoption of the euro, the franc.

However, it is also possible that the word is derived from the ethnic name of the Franks, because as the conquering class only the Franks had the status of freemen. In German, France is still called *Frankreich*, which literally means "Realm of the Franks". In order to distinguish from the Frankish Empire of Charlemagne, Modern France is called *Frankreich*, while the Frankish Realm is called *Frankenreich.*

The word "Frank" had been loosely used from the fall of Rome to the Middle Ages, yet from Hugh Capet's coronation as "King of the Franks" ("Rex Francorum") it became usual to strictly refer to the Kingdom of Francia, which would become France. The Capetian Kings were descended from the Robertines, who had produced two Frankish kings, and previously held the title of "Duke of the Franks" ("dux Francorum"). This Frankish duchy encompassed most of modern northern France but because the royal power was sapped by regional princes the term was then applied to the royal demesne as shorthand. It was finally the name adopted for the entire Kingdom as central power was affirmed over the entire kingdom.

History

Rome to Revolution

The borders of modern France are approximately the same as those of ancient Gaul, which was inhabited by Celtic *Gauls*. Gaul was conquered for Rome by Julius Caesar in the 1st century BC, and the Gauls eventually adopted Roman speech (Latin, from which the French language evolved) and Roman culture. Christianity first appeared in the 2nd and 3rd centuries AD, and became so firmly established by the fourth and fifth centuries that St. Jerome wrote that Gaul was the only region "free from heresy".

In the 4th century AD, Gaul's eastern frontier along the Rhine was overrun by Germanic tribes, principally the Franks, from whom the ancient name of "Francie" was derived. The modern name "France" derives from the name of the feudal domain of the Capetian Kings of France around Paris. The Franks were the first tribe among the Germanic conquerors of Europe after the fall of the Roman Empire to convert to Catholic Christianity rather than Arianism (their King Clovis did so in 498); thus France obtained the title "Eldest daughter of the Church" (*La fille ainee de l'Eglise*), and the French would adopt this as justification for calling themselves "the Most Christian Kingdom of France".

Existence as a separate entity began with the Treaty of Verdun (843), with the division of Charlemagne's Carolingian Empire into East Francia, Middle Francia and West Francia. Western Francia approximated the area occupied by modern France and was the precursor to modern France.

The Carolingian dynasty ruled France until 987, when Hugh Capet, Duke of France and Count of Paris, was crowned King of France. His descendants, the Direct Capetians, the House of Valois and the House of Bourbon, progressively unified the country through a series of wars and dynastic inheritance. The Albigensian Crusade was launched in 1209 to eliminate the heretical Cathars of Occitania (the south of modern-day

France). In the end, both the Cathars and the independence of southern France were exterminated. In 1066, the Duke of Normandy added King of England to his titles. Later Kings expanded their territory to cover over half of modern continental France, including most of the North, Centre and West of France. The exact boundaries changed greatly with time, but French landholdings of the English Kings remained extensive for centuries. Strong French counterattacks, helped by English weakness during the Wars of the Roses, won back mainland territory until only Calais remained. Under Mary I of England this was lost to the Spanish Netherlands.

Charles IV (The Fair) died without heir in 1328. Under the rules of the Salic Law adopted in 1316, the crown of France could not pass to a woman, nor could the line of kinship pass through the female line. Accordingly, the crown passed to the cousin of Charles, Philip of Valois, rather than passing though the female line to Charles' nephew, Edward, who would soon become Edward III of England. In the reign of Philip of Valois, the French monarchy reached the height of its medieval power. However, Philip's seat on the throne was contested by Edward III of England and in 1337, on the eve of the first wave of the Black Death, England and France went to war in what would become known as the Hundred Years' War.

In the most notorious incident during the French Wars of Religion (1562–98), thousands of Huguenots were murdered in the St. Bartholomew's Day massacre of 1572.

The monarchy reached its height during the 17th century and the reign of Louis XIV. At this time France possessed the largest population in Europe and had tremendous influence over European politics, economy, and culture. French became, and remained until the 20th century, the common language of diplomacy in international affairs. Much of the Enlightenment occurred in French intellectual circles, and major scientific breakthroughs were achieved by French scientists in the 18th century. In addition, France obtained many overseas possessions in the Americas, Africa and Asia.

Monarchy to Republic

The monarchy ruled France until the French Revolution. It did not fall immediately after the storming of the Bastille on 14 July 1789, but endured until the creation of the First Republic in September 1792. Louis XVI and his wife, Marie Antoinette, were executed (in 1793), along with thousands of other French citizens during the Reign of Terror. A guerrilla war and counterrevolution, known as the Revolt in the Vendee, cost more than 100,000 lives before it was crushed in 1796. After a series of short-lived governmental schemes, Napoleon Bonaparte seized control of the Republic in 1799, making himself First Consul, and later Emperor of what is now known as the First Empire (1804–1814). In the course of several wars, his armies conquered most of continental Europe, with members of the Bonaparte family being appointed as monarchs of newly established kingdoms. About a million Frenchmen died during the Napoleonic wars.

Following Napoleon's final defeat in 1815 at the Battle of Waterloo, the French monarchy was re-established, but with new constitutional limitations. In 1830, a civil uprising established the constitutional July Monarchy, which lasted until 1848. The short-lived Second Republic ended in 1852 when Louis-Napoléon Bonaparte proclaimed the Second Empire. Louis-Napoleon was unseated following defeat in the Franco-Prussian war of 1870 and his regime was replaced by the Third Republic.

France had colonial possessions, in various forms, since the beginning of the 17th century until the 1960s. In the 19th and 20th centuries, its global overseas colonial empire was the second largest in the world behind the British Empire. At its peak, between 1919 and 1939, the second French colonial empire extended over 12,347,000 square kilometres (4,767,000 sq mi) of land. Including metropolitan France, the total area of land under French sovereignty reached 12,898,000 square kilometres (4,980,000 sq mi) in the 1920s and 1930s, which is 8.6% of the world's land area.

France was an occupied nation in World War I and World War II. The human and material losses in the first war, which left 1.4 million French soldiers dead, exceeded largely those of the second, even though only a minor part of its territory was occupied during World War I. The interbellum phase was marked by a variety of social reforms introduced by the Popular Front government. Following the German *Blitzkrieg* campaign in World War II metropolitan France was divided in an occupation zone in the north and Vichy France, a newly established authoritarian regime collaborating with Germany, in the south.

The Fourth Republic was established after World War II and, despite spectacular economic growth (*les Trente Glorieuses*), it struggled to maintain its political status as a dominant nation state. France attempted to hold on to its colonial empire, but soon ran into trouble. The half-hearted 1946 attempt at regaining control of French Indochina resulted in the First Indochina War, which ended in French defeat at the Battle of Dien Bien Phu in 1954. Only months later, France faced a new, even harsher conflict in Algeria.

The debate over whether or not to keep control of Algeria, then home to over one million European settlers, wracked the country and nearly led to civil war. In 1958, the weak and unstable Fourth Republic gave way to the Fifth Republic, which contained a strengthened Presidency. In the latter role, Charles de Gaulle managed to keep the country together while taking steps to end the war. The Algerian War was concluded with peace negotiations in 1962 that led to Algerian independence.

In recent decades, France's reconciliation and cooperation with Germany have proved central to the political and economic integration of the evolving European Union, including the introduction of the euro in January 1999. France has been at the forefront of the European Union member states seeking to exploit the momentum of monetary union to create a more unified and capable European Union political, defence, and security apparatus. The French electorate voted against ratification of the European Constitutional Treaty in May 2005, but the successor Treaty of Lisbon was ratified by Parliament in February 2008.

Geography

France is a country in Western Europe, bordering the Atlantic Ocean (Bay of Biscay) and English Channel between Belgium and Spain.

Area

- *Total area*: 674,843 km^2

  - (Whole territory of the French Republic, including all the overseas departments and territories, but excluding the disputed French territory of Terre Adelie in Antarctica)
- *Metropolitan France:* 551,695 km^2
  - (Metropolitan-i.e. European-France only, French National Geographic Institute data)
- *Metropolitan France:* 543,965 km^2
  - (Metropolitan-i.e. European-France only, French Land Register data, which exclude lakes, ponds, glaciers larger than 1 km^2, and estuaries).

Boundaries

- Land boundaries:
  - *Total:* 4082.2 km
  - 2889 km (metropolitan), 1183 km (French Guiana) 10.2 km (Saint Martin)
- Border countries:
  - Andorra 56.6 km, Belgium 620 km, Germany 451 km, Italy 488 km, Luxembourg 73 km, Monaco 4.4 km, Spain 623 km, Switzerland 573 km (metropolitan)
  - Brazil 673 km, Suriname 510 km, French Guiana 1183 km
  - Netherlands Antilles (Sint Maarten) 10.2 km; (Saint Martin).
- Coastline: 3,427 km (metropolitan), 378 km (French Guiana), 306 km (Guadeloupe), 350 km (Martinique), 207 km (Reunion)
- Maritime claims:
  - Contiguous zone: 24 nmi (44.4 km; 27.6 mi)
  - Continental shelf: 200-m depth or to the depth of exploitation
  - Exclusive economic zone: 200 nmi (370.4 km; 230.2 mi); does not apply to the Mediterranean
  - Territorial sea: 12 nmi (22.2 km; 13.8 mi).

Internal Divisions

Metropolitan (i.e. European) France is divided into 22 *regions* (although strictly speaking Corsica is in fact a territorial collectivity, not a *region*, but is referred to as a region in common speech), which are subdivided into 96 *departements*, which are further divided into 329 arrondissements, which are further divided into 3,879 cantons, which are further divided into 36,568 communes (as of 1/1/2004). The current number is unknown.

The French Republic is further made up of the following:

Four overseas regions (*regions d'outre-mer*, or *ROM*): Guadeloupe, French Guiana, Martinique, and Reunion, which have the same status as metropolitan regions (as much as Hawaii has the same status as a continental US state), each of these overseas regions also being an overseas departement (*departement d'outre-mer*, or *DOM*), with the same status as a *departement* of metropolitan France.

This double structure (*region/departement*) is new, due to the recent extension of the regional scheme to the overseas departements, and may soon transform into a single structure, with the merger of the regional and departmental assemblies, unless new *departements* are created such as in the case of Reunion, where it has been proposed to create a second *departement* in the south of the island, with the *region* of Reunion above these two *departements*.

Five overseas collectivities (*Collectivites d'outre-mer*, or *COM*): Saint-Pierre and Miquelon, Saint Barthelemy, Saint Martin, Wallis and Futuna, and Mayotte (although strictly speaking Mayotte is in fact a "departmental collectivity", not an overseas collectivity, with the possibility to become a full-status French *departement* in 2010, but for the sake of clarity it is most often classified as overseas collectivity).

One overseas "country" (*pays d'outre-mer*, or *POM*): French Polynesia. In 2003 it became an overseas collectivity (or *COM*). Its statutory law of 27 February 2004 gives it the particular designation of overseas country inside the Republic (or *POM*), but without legal modification of its status.

One sui generis collectivity (*collectivite sui generis*): New Caledonia, whose status is unique in the French Republic.

One overseas territory (*territoire d'outre-mer*, or *TOM*): the French Southern and Antarctic Lands divided into 5 districts: Kerguelen Islands, Crozet Islands, Amsterdam Island and Saint Paul Island, Adelie Land, and the Scattered islands (Banc du Geyser, Bassas da India, Europa, Juan de Nova, Glorioso, and Tromelin).

One uninhabited island in the Pacific Ocean off the coast of Mexico which belongs directly to the central State public land and is administered by the high-commissioner of the French Republic in French Polynesia: Clipperton

Terrain

Mostly flat plains or gently rolling hills in north and west. Elevation extremes:

- Lowest point: Rhone River delta-2 m
- Highest point: Mont Blanc 4,808 m.

Natural Resources

Coal, iron ore, bauxite, fish, timber, potash, and zinc

Land Use

- *Arable land*: 33%
- *Permanent crops*: 2%
- *Permanent pastures*: 20%
- *Forests and woodland*: 27%
- *Other*: 18% (1993 est.)

Irrigated land: 16,300 km^2 (1995 est.)

Environment

Environmental Issues

Some forest damage from acid rain (major forest damage occurred as a result of severe December 1999 windstorm); air pollution from industrial and vehicle emissions; water pollution from urban wastes, agricultural runoff, high pesticide use (honeybee mortality)

International Environmental Agreements

Party to: Air Pollution, Air Pollution-Nitrogen Oxides, Air Pollution-Sulphur 85, Air Pollution-Sulphur 94, Air Pollution-

Volatile Organic Compounds, Antarctic-Environmental Protocol, Antarctic Treaty, Biodiversity, Climate Change, Desertification, Endangered Species, Hazardous Wastes, Law of the Sea, Marine Dumping, Marine Life Conservation, Ozone Layer Protection, Ship Pollution, Tropical Timber 83, Tropical Timber 94, Wetlands, Whaling

Signed, but not ratified: Air Pollution-Persistent Organic Pollutants, Climate Change-Kyoto Protocol

Picture from Space

This image of metropolitan France was generated with data from the Shuttle Radar Topography Mission (SRTM). For this broad view the resolution of the data was reduced, resampled to a Mercator projection and the French border outlined. The variety of landforms comprising the country is readily apparent.

The upper central part of this scene is dominated by the Paris Basin, which consists of a layered sequence of sedimentary rocks. Fertile soils over much of the area make good agricultural land. The Normandie coast to the upper left is characterized by high, chalk cliffs, while the Brittany coast (the peninsula to the left) is highly indented where deep valleys were drowned by the sea, and the Biscay coast to the southwest is marked by flat, sandy beaches.

To the south, the Pyrenees form a natural border between France and Spain, and the south-central part of the country is dominated by the ancient Massif Central. Subject to volcanism that has only subsided in the last 10,000 years, these central mountains are separated from the Alps by the north-south trending Rhone River Basin.

Two visualization methods were combined to produce the image: shading and colour coding of topographic height. The shade image was derived by computing topographic slope in the northwest-southeast direction, so that northwest slopes appear bright and southeast slopes appear dark. Colour coding is directly related to topographic height, with green at the lower elevations, rising through yellow and tan, to white at the highest elevations.

Elevation data used in this image were acquired by the Shuttle Radar Topography Mission aboard the Space Shuttle

Endeavour, launched on Feb. 11, 2000. SRTM used the same radar instrument that comprised the Spaceborne Imaging Radar-C/X-band Synthetic Aperture Radar (SIR-C/X-SAR) that flew twice on the Space Shuttle Endeavour in 1994. SRTM was designed to collect 3-D measurements of the Earth's surface. To collect the 3-D data, engineers added a 60-meter (approximately 200-foot) mast, installed additional C-band and X-band antennas, and improved tracking and navigation devices.

- Location: 42 to 51.5° north, 5.5 west to 8° east.
- Orientation: north toward the top, Mercator projection.
- Image Data: shaded and coloured SRTM elevation model.
- Original Data Resolution: 1 arcsecond (~30 meters).
- Date Acquired: February 2000.
- Image Courtesy SRTM Team NASA/JPL/NIMA.

Extreme Points

This is a list of the extreme points of France; the points that are farther north, south, east or west than any other location.

France (Metropolitan)

- Northernmost Point — Bray-Dunes, Nord at 51°05'N 2°32'E/51.083°N 2.533°
- Southernmost Point — Iles Lavezzi, off Corsica at 41°20N 9°15'E/41.333°N 9.25°E
- Westernmost Point — Ile d'Ouessant, off Brittany at 48°27'N 5°08'W/48.45°N 5.133°W
- Easternmost Point — near Cervione, Haute-Corse at 42°17'N 9°33'E/42.283°N 9.55°E.

France (Mainland)

- Northernmost Point — Bray-Dunes, Nord at 51°05'N 2°32'E/51.083°N 2.533°E
- Southernmost Point — Puig de Comanegra, Pyrenees-Orientales at 42°20'N 2°31'E/42.333°N 2.517°E
- Westernmost Point — Pointe de Corsen, Finistere at 48°24'N 4°47'W/48.4°N 4.783°W
- Easternmost Point — Lauterbourg, Bas Rhin at 48°58' N 8°13'E/48.967°N 8.217°E.

France (Including Departements D'outre Mer*)*

- Northernmost Point — Bray-Dunes, Nord at 51°05'N 2°32'E/51.083°N 2.533°E
- Southernmost Point — Saint-Joseph, Reunion at 22°23' S 55°38'E/22.383°S 55.633°E
- Westernmost Point — La Pointe-Noire, Guadeloupe at 16°16'N 61°48'W/16.267°N 61.8°W
- Easternmost Point — Sainte-Rose, Réunion at 21°11'S 55°50'E/21.183°S 55.833°E.

France (Territory of the French Republic, Including Collectivites Territorials *and* Pays et Territories D'outre-mer*)*

- Northernmost Point : Bray-Dunes, Nord at 51°05'N 2°32'E/51.083°N 2.533°E
- Southernmost Point : Iles de Boynes, Kerguelen Islands, French Southern and Antarctic Lands at 50°01'S 68°52'E/ 50.017°S 68.867°E
- Westernmost Point : Toloke, Futuna, Wallis and Futuna at 14°42'S 178°33'W/14.7°S 178.55°W
- Easternmost Point : Hunter Island, New Caledonia at 22°31'S 172°62 E.

While Metropolitan France is located in Western Europe, France also has a number of territories in North America, the Caribbean, South America, the southern Indian Ocean, the Pacific Ocean, and Antarctica. These territories have varying forms of government ranging from overseas department to overseas collectivity.

Metropolitan France covers 547,030 square kilometres (211,209 sq mi), having the largest area among European Union members and slightly larger than Spain. France possesses a wide variety of landscapes, from coastal plains in the north and west to mountain ranges of the Alps in the south-east, the Massif Central in the south-central and Pyrenees in the south-west. At 4,810.45 metres (15,782 ft) above sea-level, the highest point in Europe, Mont Blanc, is situated in the Alps on the border between France and Italy. Metropolitan France also has extensive river systems such as the Seine, the Loire, the Garonne, and the Rhone, which divides the Massif Central from the Alps and flows into the Mediterranean Sea at the Camargue, the lowest point in France (2 m/6.56 ft below sea level). Corsica lies off the Mediterranean coast.

France's total land area, with its overseas departments and territories (excluding Adelie Land), is 674,843 km^2 (260,558 sq mi), 0.45% of the total land area on Earth. However, France possesses the second-largest Exclusive Economic Zone (EEZ) in the world, covering 11,035,000 km^2 (4,260,637 sq mi), approximately 8% of the total surface of all the EEZs of the world, just behind the United States (11,351,000 km^2/ 4,382,646 sq mi) and ahead of Australia (8,232,000 km^2/ 3,178,393 sq mi).

Metropolitan France is situated between 41° and 51° North, on the western edge of Europe, and thus lies within the northern temperate zone. The north and northwest have a temperate climate, while a combination of maritime influences, latitude and altitude produce a varied climate in the rest of Metropolitan France. In the south-east a Mediterranean climate prevails. In the west, the climate is predominantly oceanic with a high level of rainfall, mild winters and cool to warm summers. Inland the climate becomes more continental with hot, stormy summers, colder winters and less rain. The climate of the Alps and other mountainous regions is mainly alpine, with the number of days with temperatures below freezing over 150 per year and snow cover lasting for up to six months.

Cities

The largest cities in France, in terms of metropolitan area population, are Paris (11,769,433), Lyon (1,748,271), Marseille (1,605,000), Lille (1,164,716), Nice (1,197,751), Toulouse (1,102,882), Bordeaux (999,149) and Nantes (804,000).

Government

The French Republic is a unitary semi-presidential republic with strong democratic traditions. The constitution of the Fifth Republic was approved by referendum on 28 September 1958. It greatly strengthened the authority of the executive in relation to parliament. The executive branch itself has two leaders: the President of the Republic, currently Nicolas Sarkozy, who is head of state and is elected directly by universal adult suffrage for a 5-year term (formerly 7 years), and the Government, led by the president-appointed Prime Minister, currently Francois Fillon.

The French parliament is a bicameral legislature comprising a National Assembly (*Assemblee Nationale*) and a Senate. The National Assembly deputies represent local constituencies and are directly elected for 5-year terms. The Assembly has the power to dismiss the cabinet, and thus the majority in the Assembly determines the choice of government. Senators are chosen by an electoral college for 6-year terms (originally 9-year terms), and one half of the seats are submitted to election every 3 years starting in September 2008.

The Senate's legislative powers are limited; in the event of disagreement between the two chambers, the National Assembly has the final say, except for constitutional laws and *lois organiques* (laws that are directly provided for by the constitution) in some cases. The government has a strong influence in shaping the agenda of Parliament.

French politics are characterised by two politically opposed groupings: one left-wing, centred around the French Socialist Party, and the other right-wing, centred previously around the Rassemblement pour la Republique (RPR) and now its successor the Union for a Popular Movement (UMP). The executive branch is currently composed mostly of the UMP.

Conventions and Notations

France is the home of the International System of Units (the metric system). Some pre-metric units are still used, essentially the *livre* (a unit of weight equal to half a kilogram) and the *quintal* (a unit of weight equal to 100 kilograms).

In mathematics, France uses the infix notation like most countries. For large numbers the long scale is used. Thus, the French use the word *billion* for the number 1,000,000,000,000, which in countries using short scale is called a trillion. However, there exists a French word, *milliard*, for the number 1,000,000,000, which in countries using the short scale is called a billion. Thus, despite the use of the long scale, one billion is called *un milliard* ("one milliard") in French, and not *mille millions* ("one thousand million"). It should also be noted that names of numbers above the *milliard* are rarely used. Thus, one trillion will most often be called *mille milliards* ("one thousand milliard") in French, and rarely *un billion*. In the French numeral notation, the comma (,) is the decimal separator, whereas a space is used between each group of three digits (fifteen million five hundred thousand and thirty-two should be written as 15 500 032). In finance, the currency symbol is used as a decimal separator or put after the number. For example, €25,048.05 is written either 25 048€05 or 25 048,05 € (always with an extra space between the figure and the currency symbol).

In computing, a bit is called a bit yet a byte is called an octet (from the Latin root *octo*, meaning "8"). SI prefixes are used.

24-hour clock time is used, with h being the separator between hours and minutes (for example 2:30 p.m. is 14h30).

The all-numeric form for dates is in the order day-month-year, using a slash as the separato.

Law

France uses a civil legal system; that is, law arises primarily from written statutes; judges are not to make law, but merely to interpret it (though the amount of judge interpretation in certain areas makes it equivalent to case law). Basic principles of the rule of law were laid in the Napoleonic Code. In agreement with the principles of the Declaration of the Rights of Man and of the Citizen law should only prohibit actions detrimental to society. As Guy Canivet, first president of the Court of Cassation, wrote about the management of prisons:

Freedom is the rule, and its restriction is the exception; any restriction of Freedom must be provided for by Law and must follow the principles of necessity and proportionality.

That is, Law should lay out prohibitions only if they are needed, and if the inconveniences caused by this restriction do not exceed the inconveniences that the prohibition is supposed to remedy.

French law is divided into two principal areas: private law and public law. Private law includes, in particular, civil law and criminal law. Public law includes, in particular, administrative law and constitutional law. However, in practical terms, French law comprises three principal areas of law: civil law, criminal law and administrative law.

France does not recognise religious law, nor does it recognise religious beliefs or morality as a motivation for the enactment of prohibitions. As a consequence, France has long had neither blasphemy laws nor sodomy laws (the latter being abolished in 1791). However "offences against public decency" or disturbing public order (*trouble a l'ordre public*) have been used to repress public expressions of homosexuality or street prostitution.

Criminal laws can only address the future and not the past (criminal *ex post facto* laws are prohibited) ; and to be applicable, laws must be officially published in the *Journal Official de la Republique Francaise*.

Foreign Relations

France is a member of the United Nations and serves as one of the permanent members of the U.N. Security Council with veto rights. It is also a member of the World Trade Organisation (WTO), the Secretariat of the Pacific Community (SPC) and the Indian Ocean Commission (COI). It is an associate member of the Association of Caribbean States (ACS) and a leading member of the International Francophone Organisation (OIF) of fifty-one fully or partly French-speaking countries. It hosts the headquarters of the OECD, UNESCO, Interpol, Alliance Base and the International Bureau for Weights and Measures. In 1953 France received a request from the United Nations to pick a coat of arms that would represent it internationally. Thus the French emblem was adopted and is currently used on passports.

French foreign policy has been largely shaped by membership of the European Union, of which it was a founding member. In the 1960s, France sought to exclude the British from the organisation, seeking to build its own standing in continental Europe. Since the 1990s, France has developed close ties with reunified Germany to become the most influential driving force of the EU, but consequently rivalling the UK and limiting the influence of newly inducted East European nations.

France is a member of the North Atlantic Treaty Organisation, but under President de Gaulle, it excluded itself from the joint military command to avoid the supposed domination of its foreign and security policies by US political and military influence. However, as a result of Nicolas Sarkozy's (much criticized in France) *pro-American* politics, France rejoined the NATO joint military command on 4 April 2009. In the early 1990s, the country drew considerable criticism from other nations for its underground nuclear tests in French Polynesia. France vigorously opposed the 2003 invasion of Iraq, straining bilateral relations with the US and the UK. France retains strong political and economic influence in its former African colonies (Francafrique) and has supplied economic aid and troops for peace-keeping missions in the Ivory Coast and Chad.

Military

The French armed forces are divided into four branches:

- Armee de Terre (Army)
- Marine Nationale (Navy)
- Armee de l'Air (Air Force)
- Gendarmerie Nationale (A military force which acts as a National Rural Police and as a Military police for the entire French military).

After the Algerian War, conscription was steadily reduced and was finally suspended in 2001 by President Jacques Chirac. The total number of military personnel is approximately 359,000. France spends 2.6% of its GDP on defence, slightly more than the United Kingdom (2.4%) and the highest in the European Union where defence spending generally accounts to less than 1.5% of GDP. France and the U.K. account for 40% of EU defence spending. About 10% of France's defence budget goes towards its nuclear deterrence, or nuclear weapons force.

France has major military industries that have produced the Rafale fighter, the Charles de Gaulle aircraft carrier, the Exocet missile and the Leclerc tank amongst others. Some weaponry, like the E-2 Hawkeye or the E-3 Sentry was bought from the United States. Despite withdrawing from the Eurofighter project, France is actively investing in European joint projects such as the Eurocopter Tiger, multipurpose frigates, the UCAV demonstrator nEUROn and the Airbus A400M.

France is a major arms seller as most of its arsenal's designs are available for the export market with the notable exception of nuclear-powered devices. Some of the French designed equipments are specifically designed for exports like the Franco-Spanish *Scorpene* class submarines. Some French equipments have been largely modified to fit allied countries' requirements like the *Formidable* class frigates (based on the *La Fayette* class) or the *Hashmat* class submarines (based on the *Agosta* class submarines).

Although it includes very competent anti-terrorist units such as the GIGN or the EPIGN, the gendarmerie is a military police force which serves for the most part as a rural and general purpose police force. Since its creation the GIGN has taken part in roughly one thousand operations and freed over five-hundred hostages; the Air France Flight 8969's hijacking brought them to the world's attention.

French intelligence constitutes of two major units: the DGSE (the external agency) and the DCRI (domestic agency). The latter being part of the police while the former is associated to the army. The DGSE is notorious for the Sinking of the *Rainbow Warrior*, but it is also known for revealing the most extensive technological spy network uncovered in Europe and the United States to date through the mole Vladimir Vetrov.

The French deterrence, (formerly known as "*Force de frappe*"), relies on a complete independence. The current French nuclear force consists of four submarines equipped with M45 ballistic missiles. The current *Triomphant* class is currently under deployment to replace the former *Redoutable* class. The M51 will replace the M45 in the future and expand the *Triomphants* firing range. Aside of the submarines the French dissuasion force uses the Mirage 2000N; it is a variant of the Mirage 2000 and thus is designed to deliver nuclear strikes. Other nuclear devices like the Plateau d'Albion's Intermediate-range ballistic missile and the short range Hadès missiles have been disarmed. With 350 nuclear heads stockpiled France is the world's third largest nuclear power.

The Marine Nationale is regarded as one of the world's most powerful navies. The professional compendium *flottes de combats*, in its 2006 edition, ranked it world's 6th biggest navy after the American, Russian, Chinese, British and Japanese navies. It is equipped with the only non-American nuclear powered Aircraft Carrier in the world. Recently *Mistral* class ships joined the Marine Nationale, the Mistral itself having taken part to operations in Lebanon. For the 2004 centennial of the Entente Cordiale President Chirac announced the Future French aircraft carrier would be jointly designed with Great Britain. The French navy is equipped with the *La Fayette* class frigates, early examples of stealth ships, and several ships are expected to be retired in the next few years and replaced by more modern ships, examples of future surface ships are the Forbin and the *Aquitaine* class frigates. The attack submarines are also part of the Force Oceanique Strategique although they do not carry the nuclear dissuasion, the current class is the *Rubis* Class and will be replaced in the future by the expected *Suffren* Class.

The *Armee de Terre* employs 133,500 people. It is famous for the *Legion Etrangere* (French Foreign Legion) though the French special forces are not the Legion but the Dragons Parachutistes and the Marines Parachutistes. The French assault rifle is the FAMAS and future infantry combat system is the Felin. France uses both tracked and wheeled vehicles to a significant points, examples of wheeled vehicles would be the Caesar or the AMX 10 RC. Although its main battle tank is the Leclerc many older AMX 30 tanks are still operational. It uses the AMX 30 AuF1 for artillery and is equipped with Eurocopter Tigers helicopters.

The Armee de l'Air is the oldest and first professional air force worldwide. It still today retains a significant capacity. It uses mainly two aircraft fighters: the older Mirage F1 and the more recent Mirage 2000. The later model exists in a ground attack version called the Mirage 2000D. The modern Rafale is in deployment in both the French air force and navy.

Transport

The railway network of France, which stretches 31,840 kilometres (19,784 mi) is the most extensive in Western Europe. It is operated by the SNCF, and high-speed trains include the Thalys, the Eurostar and TGV, which travels at 320 km/h (199 mph) in commercial use. The Eurostar, along with the Eurotunnel Shuttle, connects with the United Kingdom through the Channel Tunnel. Rail connections exist to all other neighbouring countries in Europe, except Andorra. Intra-urban connections are also well developed with both underground services and tramway services complementing bus services. There are approximately 893,300 kilometres (555,071 mi) of serviceable roadway in France. The Paris region is enveloped with the most dense network of roads and highways that connect it with virtually all parts of the country. French roads also handle substantial international traffic, connecting with cities in neighboring Belgium, Spain, Andorra, Monaco, Switzerland, Germany and Italy. There is no annual registration fee or road tax; however, motorway usage is through tolls except in the vicinity of large communes. The new car market is dominated by domestic brands such as Renault (27% of cars sold in France in 2003), Peugeot (20.1%) and Citroën (13.5%). Over 70% of new cars sold in 2004 had diesel engines, far more than contained petrol or LPG engines. France possesses the world's tallest road bridge: the Millau Viaduct, and has built many important bridges such as the Pont de Normandie.

Marine Transport

France has an extensive merchant marine, including 55 ships of size GRT | 1,000 and above. The country also maintains a captive register for French-owned ships in Iles Kerguelen (French Southern and Antarctic Lands).

French companies operate over 1,400 ships of which 700 are registered in France. France's 110 shipping firms employ 12,500 personnel at sea and 15,500 on shore. Each year, 305 million tonnes of goods and 15 million passengers are transported by sea. Marine transport is responsible for 72% of France's imports and exports.

France also boasts a number of seaports and harbours, including Bayonne, Bordeaux, Boulogne-sur-Mer, Brest, Calais, Cherbourg-Octeville, Dunkerque, Fos-sur-Mer, La Pallice, Le Havre, Lorient, Marseille, Nantes, Nice, Paris, Port-la-Nouvelle, Port-Vendres, Roscoff, Rouen, Saint-Nazaire, Saint-Malo, Sete, Strasbourg and Toulon.

Air Travel

There are approximately 478 airports in France (1999 est.) and by a 2005 estimate, there are three heliports. 288 of the airports have paved runways, with the remaining 199 being unpaved.

Among the airspace governance authorities active in France, one is Aeroports de Paris, which has authority over the Paris region, managing 14 airports including the two busiest in France, Charles de Gaulle Airport and Orly Airport. The former, located in Roissy near Paris, is the fifth busiest airport in the world with 60 million passenger movements in 2008, and France's primary international airport, serving over 100 airlines. The national carrier of France is Air France, a full service global airline which flies to 20 domestic destinations and 150 international destinations in 83 countries (including Overseas departments and territories of France) across all 6 major continents.

Administrative Divisions

France is divided into 26 administrative regions. 22 are in metropolitan France (21 are on the continental part of metropolitan France; one is the territorial collectivity of Corsica), and four are overseas regions. The regions are further subdivided into 100 departments which are numbered (mainly alphabetically). This number is used in postal codes and vehicle number plates amongst others. The 100 departments are subdivided into 341 arrondissements which are, in turn, subdivided into 4,032 cantons. These cantons are then divided into 36,680 communes, which are municipalities with an elected municipal council. There also exist 2,588 inter communal entities grouping 33,414 of the 36,680 communes (i.e. 91.1% of all the communes). Three communes, Paris, Lyon and Marseille are also subdivided into 45 municipal arrondissements. The regions, departments and communes are all known as territorial collectivities, meaning they possess local assemblies as well as an executive. Arrondissements and cantons are merely administrative divisions. However, this was not always the case. Until 1940, the arrondissements were also territorial collectivities with an elected assembly, but these were suspended by the Vichy regime and definitely abolished by the Fourth Republic in 1946. Historically, the cantons were also territorial collectivities with their elected assemblies.

Overseas Regions/Departments, Collectivities and Territories

Among the 100 departments of France, four (French Guiana,

Guadeloupe, Martinique, and Réunion) are in overseas regions (ROMs) that are also simultaneously overseas departments (DOMs) and are an integral part of France (and the European Union) and thus enjoy a status similar to metropolitan departments.

In addition to the 26 regions and 100 departments, the French Republic also has six overseas collectivities (French Polynesia, Mayotte, Saint Barthelemy, Saint Martin, Saint Pierre and Miquelon, and Wallis and Futuna), one *sui generis* collectivity (New Caledonia), one overseas territory (French Southern and Antarctic Lands), and one island possession in the Pacific Ocean (Clipperton Island).

Overseas collectivities and territories form part of the French Republic, but do not form part of the European Union or its fiscal area. The Pacific Collectivities (COMs) of French Polynesia, Wallis and Fortuna, and New Caledonia continue to use the Pacific franc whose value is linked to that of the euro. In contrast, the four overseas regions used the French franc and now use the euro.

Economy

The economy of France is a mixed economy which is the fifth largest in the world in nominal terms, behind the United States, Japan, China and Germany and the eighth largest by purchasing power parity. It is the second largest economy in Europe behind Germany in nominal terms and fourth largest behind Germany, United Kingdom and Russia by PPP. On May 15, 2009, the INSEE announced that France has officially entered a recession after its GDP decreased by 1.2% of Q1 in 2009. Rise and Decline of Dirigisme

France embarked on an ambitious and very successful programme of modernization under state impulse and coordination. This program of *dirigisme*, mostly implemented by left-wing governments, involved the state control of certain industries such as transportation, energy and telecommunication infrastructures as well as various incentives for private corporations to merge or engage in certain projects.

However, *dirigisme* came to be highly contested after 1982 when newly elected socialist president Francois Mitterrand called for increased governmental control in the economy, nationalising many industries and private banks. By 1983, the government decided to renounce *dirigisme* and start the era of rigueur ("rigour") or corporatization. As a result the government largely retreated from economic intervention; *dirigisme* has now essentially receded though some of its traits remain.

Despite significant liberalisation over the past 15 years, the government continues to play a significant role in the economy: government spending, at 53% of GDP in 2001, is the highest in the G-7. Labour conditions and wages are highly regulated. The government continues to own shares in corporations in a range of sectors, including banking, energy production and distribution, automobiles, transportation, and telecommunications which differs from countries like the U.S or U.K where most of these companies are privatised.

Sectors of the Economy

Leading industrial sectors in France are telecommunications (including communication satellites), aerospace and defence, ship building (naval and specialist ships), pharmaceuticals, construction and civil engineering, chemicals, and automobile production (3.5m units in 2005).

Research and development spending is also high in France at 2.26% of GDP, the fourth highest in the OECD.

Energy

With no domestic oil production, France has relied heavily on the development of nuclear power, which now accounts for about 78% of the country's electricity production, up from only 8% in 1973, 24% in 1980, and 75% in 1990. Nuclear waste is stored on site at reprocessing facilities.

In 2006 of electricity in France amounted to 548.8 TWh, of which:

- 428.7 TWh (78.1%) were produced by nuclear power generation
- 60.9 TWh (11.1%) were produced by hydroelectric power generation
- 52.4 TWh (9.5%) were produced by fossil fuel power generation
    - 21.6 TWh (3.9%) by coal power

- o 20.9 TWh (1.1%) by natural gas power
- o 9.9 TWh (1.8%) by other fossil fuel generation (fuel oil and gases by-products of industry such as blast furnace gases)

- 6.9 TWh (1.3%) were produced by other types of power generation (essentially waste-to-energy and wind turbines))
  - o The electricity produced by wind turbines increased from 0.596 TWh in 2004, to 0.963 TWh in 2005, and 2.15 TWh in 2006, but this still accounts only for 0.4% of the total production of electricity (as of 2006).

Inflation rate 2003 1.80% 2004 2.10% 2005 2.30% 2006 1.70% 2007 1.50% 2008 1.50% 2009 2.80%

Privatisation of EDF

In November 2004, EDF (which stands for Electricity de France), the largest electricity provider in France, was floated on the French stock market, with the French State keeping more than 70% of the capital. EDF is not the only electricity provider in France. Other electricity providers include CNR (*Compagnie nationale du Rhone*) and Endesa (through SNET).

Agriculture

France is the European Union's leading agricultural producer, accounting for about one-third of all agricultural land within the EU. Northern France is characterized by large wheat farms. Dairy products, pork, poultry, and apple production are concentrated in the western region. Beef production is located in central France, while the production of fruits, vegetables, and wine ranges from central to southern France. France is a large producer of many agricultural products and is currently expanding its forestry and fishery industries. The implementation of the Common Agricultural Policy (CAP) and the Uruguay Round of the General Agreement on Tariffs and Trade (GATT) have resulted in reforms in the agricultural sector of the economy.

France is the world's sixth-largest agricultural producer and the second-largest agricultural exporter, after the United States. However, the destination of 70% of its exports are other

EU member states and many poor African countries (including its former colonies) which face serious food shortage. Wheat, beef, pork, poultry, and dairy products are the principal exports. The United States, although the second-largest exporter to France, faces stiff competition from domestic production, other EU member states, and third world countries. U.S. agricultural exports to France, totalling some $600 million annually, consist primarily of soybeans and products, feeds and fodders, seafood, and consumer oriented products, especially snack foods and nuts. French exports to the United States are mainly cheese, processed products and wine.

The French agricultural sector is heavily dependent upon subsidies from the European Union, which account for €11 billion. France is the main country in the EU that is against the reduction of subsidies. Subsidies have given France a competitive advantage which also demotes the concept of free trade. Specific government policies, such as the infamous reclassification of French wine as a 'health food' to avoid VAT, also goes a long way to create a thriving domestic sector.

Tourism

With 81.9 million foreign tourists in 2007, France is ranked as the first tourist destination in the world, ahead of Spain (58.5 million in 2006) and the United States (51.1 million in 2006). This 81.9 million figure excludes people staying less than 24 hours in France, such as northern Europeans crossing France on their way to Spain or Italy during the summer. France features cities of high cultural interest (Paris being the foremost), beaches and seaside resorts, ski resorts, and rural regions that many enjoy for their beauty and tranquillity (green tourism). France also attracts many religious pilgrims to Lourdes, a town in the Hautes-Pyrenees departement, that hosts a few million visitors a year.

Popular tourist sites include: (according to a 2003 ranking visitors per year): Eiffel Tower (6.2 million), Louvre Museum (5.7 million), Palace of Versailles (2.8 million), Musee d'Orsay (2.1 million), Arc de Triomphe (1.2 million), Centre Pompidou (1.2 million), Mont-Saint-Michel (1 million), Chateau de Chambord (711,000), Sainte-Chapelle (683,000), Chateau du Haut-Kingsbourg (549,000), Puy de Dome (500,000), Musee Picasso (441,000), Carcassonne (362,000).

Weapons Industry

France is the fourth largest weapons supplier in the world. The French arms industry's main customer, for whom they mainly build warships, guns, nuclear weapons and equipment, is the French Government. Furthermore, record high defense expenditure (currently at €35 billion), which was considerably increased under the government of Prime Minister Jean-Pierre Raffarin, have contributed to the success of the French arms industries. In addition, external demand plays a big part in the growth of this sector: for example, France exports great quantities of weaponry to the United Arab Emirates, Brazil, Greece, India, Pakistan, Taiwan, Singapore and many others.

External Trade

France is the second-largest trading nation in Europe (after Germany). Its foreign trade balance for goods had been in surplus from 1992 until 2001, reaching $25.4 billion (25.4 G$) in 1998; however, the French balance of trade was hit by the economic downturn, and went into the red in 2000, reaching US$15bn in deficit in 2003. Total trade for 1998 amounted to $730 billion, or 50% of GDP—imports plus exports of goods and services. Trade with European Union countries accounts for 60% of French trade.

In 1998, U.S.-France trade totalled about $47 billion—goods only. According to French trade data, U.S. exports accounted for 8.7%—about $25 billion—of France's total imports. U.S. industrial chemicals, aircraft and engines, electronic components, telecommunications, computer software, computers and peripherals, analytical and scientific instrumentation, medical instruments and supplies, broadcasting equipment, and programming and franchising are particularly attractive to French importers.

The principal French exports to the United States are aircraft and engines, beverages, electrical equipment, chemicals, cosmetics, luxury products and perfume. France is the ninth-largest trading partner of the U.S.

Regions *Economy*

The economic disparity between French regions is not as high as that in other European countries such as Spain, Italy or Germany. The most important *regions* are Ile-de-France (4th agglomeration for her economy in the world), Rhones-Alpes (6th economic European region thanks to its services, high-technologies, chemical industries, wines, tourism), Provence-Alpes-Cotes d'Azur (services, industry, tourism and wines), Nord-Pas-de-Calais (industries) and Pays de la Loire.

Regions like Alsace, which has a rich past in industry (machine tool), are relatively wealthy without ranking very high in absolute term.

The rurals area are mainly in Auvergne, Limousin, and Centre, and wines productions account for a significant amount of the economy in Aquitaine (Bordeaux *region*), and champagne for Champagne-Ardennes. List of French regions ranked by GDP total and per capita.

Departements Economy and Cities

In terms of income, important inequalities can be observed among the French "departements". According to the 2008 statistics of the INSEE, the Yvelines is the richest *departement* of the country with an average household income of €3,750 per month. Hauts-de-Seine comes second, Essonne third, Paris fourth, Seine-et Marne fifth.

Ile-de-France is clearly the richest region in the country with an average household income of €3,228 per month compared to €2 478 on a national level. Alsace comes second, Rhone-Alpes third, Picardie fourth, Haute-Normandie fifth.

On the contrary, French overseas territories are the poorest parts of France, Guyanne being the poorest "departement" with an average household income of €1,826.

In metropolitan France, it is Creuse, in the Limousin region, which comes bottom of the list with an average household income of €1,849 per month.

For instance, in the Paris metropolitan area itself we can find huge differences between the standard of living of some posh western suburbs and some disadvantaged areas in the north of Paris.

If we look at cities of over 50,000 inhabitants, Neuilly-sur-Seine, a western suburb of Paris, is clearly the most prosperous city in France with an average household income of €5,434, and

20% earning more than €8,000 per month. In Paris three districts have comparable level of incomes per household : the 7th, the 8th and the 16th "arrondissement". The 8th "arrondissement" is the richest district in Paris and is even surpassing Neuilly-sur-Seine in terms of household incomes.

If we take Aubervilliers, or La Courneuve for instance, incomes are among the poorest of France. In Aubervilliers, average householf income is about €1,840 per month. The 2005 riots took place in this part of the Paris metropolitan area.

According to Boston Consulting Group, there were also 280,000 US dollar millionaires living in France in 2010, or about 0.45% of the total population.

A member of the G8 group of leading industrialised countries, it is ranked as the fifth largest economy by nominal GDP. France joined 11 other EU members to launch the euro on 1 January 1999, with euro coins and banknotes completely replacing the French franc (£) in early 2002.

France's economy combines extensive private enterprise (nearly 2.5 million companies registered) with substantial (though declining) state enterprise and government intervention. The government retains considerable influence over key segments of infrastructure sectors, with majority ownership of railway, electricity, aircraft, nuclear power and telecommunications. It has been gradually relaxing its control over these sectors since the early 1990s. The government is slowly corporatizing the state sector and selling off holdings in France Telecom, Air France, as well as the insurance, banking, and defence industries. France has an important aerospace industry led by the European consortium Airbus, and has its own national spaceport, the *Centre Spatial Guyanais*.

According to the WTO, in 2009 France was the world's sixth-largest exporter and the fourth-largest importer of manufactured goods. In 2008, France was the third-largest recipient of foreign direct investment among OECD countries at $117.9 billion, ranking behind Luxembourg (where foreign direct investment was essentially monetary transfers to banks located in that country) and the United States ($316.1 billion), but above the United Kingdom ($96.9 billion), Germany ($24.9 billion), or Japan ($24.4 billion). In the same year, French companies invested $220 billion outside of France, ranking

France as the second most important outward direct investor in the OECD, behind the United States ($311.8 billion), and ahead of the United Kingdom ($111.4 billion), Japan ($128 billion) and Germany ($156.5 billion).

France is the smallest emitter of carbon dioxide among the seven most industrialized countries in the world, due to its heavy investment in nuclear power. As a result of large investments in nuclear technology, most of the electricity produced in the country is generated by 59 nuclear power plants (78% in 2006, up from only 8% in 1973, 24% in 1980, and 75% in 1990). In this context, renewable energies are having difficulties taking off the ground.

Large tracts of fertile land, the application of modern technology, and EU subsidies have combined to make France the leading agricultural producer and exporter in Europe. Wheat, poultry, dairy, beef, and pork, as well as an internationally recognized foodstuff and wine industry are primary French agricultural exports. EU agriculture subsidies to France have decreased for the last years, but still amounted to $8 billion in 2007.

Since the end of the Second World War the government made efforts to integrate more and more with Germany, both economically and politically. Today the two countries form what is often referred to as the "core" countries in favour of greater integration of the European Union.

Labour Market

The French GDP per capita is similar the GDP per capita of other comparable European countries such as Germany and the United Kingdom. GDP per capita is determined by (i) productivity per hour worked, which in France is the highest of the G8 countries in 2005, according to the OECD, (ii) the number of hours worked, which is one the lowest of developed countries, and (iii) the employment rate. France has one of the lowest 15–64 years employment rates of the OECD countries: in 2004, only 69% of the French population aged 15–64 years were in employment, compared to 80% in Japan, 79% in the UK, 77% in the US, and 71% in Germany.

This gap is due to the very low employment rates at both age extremes: the employment rate of people aged 55–64 was

38,3% in 2007, compared to 46,6% in the EU15; for the 15–24 years old, the employment rate was 31,5% in 2007, compared to 37,2% in EU25. These low employment rates are explained by the high minimum wages which prevent low productivity workers – such as young people – from easily entering the labour market, ineffective university curricula that fail to prepare students adequately for the labour market, and, concerning the older workers, restrictive legislation on work and incentives for premature retirement.

The unemployment rate decreased from 9% in 2006 to 7% in 2008 but remains one of the highest in Europe. In June 2009, the unemployment rate for France was 9.4%. Shorter working hours and the reluctance to reform the labour market are mentioned as weak spots of the French economy in the view of the right, when the left mentions the lack of government policies fostering social justice. Liberal economists have stressed repeatedly over the years that the main issue of the French economy is an issue of structural reforms, in order to increase the size of the working population in the overall population, reduce the taxes' level and the administrative burden.

Keynesian economists have different answers to the unemployment issue, and their theories led to the 35-hour workweek law in the early 2000s, which turned out to be failure in reducing unemployment. Afterwards, between 2004 and 2008, the Government made some supply-oriented reforms to combat unemployment but met with fierce resistance, especially with the *contract nouvelle embauche* and the *contract premiere embauche* which both were eventually repealed. The current Government is experiencing the *Revenu de solidarity active.*

Demography

With an estimated population of 65.4 million people (as of 1 Jan. 2010), France is the 20th most populous country in the world. In 2003, France's natural population growth (excluding immigration) was responsible for almost all natural population growth in the European Union. In 2004, population growth was 0.68% and then in 2005 birth and fertility rates continued to increase. The natural increase of births over deaths rose to 299,800 in 2006. The total fertility rate rose to 2.02 in 2008, from 1.88 in 2002.

In 2004, a total of 140,033 people immigrated to France. Of them, 90,250 were from Africa and 13,710 from Europe. In 2005, immigration level fell slightly to 135,890.

It is illegal for the French state to collect data on ethnicity and race, a law with its origins in the 1789 revolution and reaffirmed in the constitution of 1958. Nonetheless, France is an ethnically diverse nation with about six million North Africans and an estimated 2.5 million blacks. It is currently estimated that 40% of the French population descends at least partially from the different waves of immigration the country has received. Between 1921 and 1935 about 1.1 million net immigrants came to France. An estimated 1.6 million European *pieds noirs* returned to France as the country's North African possessions gained independence.

According to the French National Institute for Statistics and Economic Studies, it has an estimated 4.9 million foreign-born immigrants, of which 2 million have acquired French citizenship. France is the leading asylum destination in Western Europe with an estimated 50,000 applications in 2005 (a 15% decrease from 2004). The European Union allows free movement between the member states. While UK and Ireland did not impose restrictions, France put in place controls to curb Eastern European migration. A perennial political issue concerns rural depopulation. Over the period 1960–1999 fifteen rural *departements* experienced a decline in population. In the most extreme case, the population of Creuse fell by 24%.

According to Article 2 of the Constitution, amended in 1992, French is the sole official language of France. Therefore, France is the only Western European nation (excluding microstates) to have only one officially recognised language. However, 77 regional languages are also spoken, in metropolitan France as well as in the overseas departments and territories. Until recently, the French government and state school system discouraged the use of any of these languages, but they are now taught to varying degrees at some schools. Other languages, such as Portuguese, Italian, Maghrebi Arabic and several Berber languages are spoken by immigrants.

Religion

Christianity is the largest religion in France, which is a

secular country, and freedom of religion is a constitutional right. According to a January 2007 poll by the Catholic World News: 60% identified as being Catholics, 31% identified as being agnostics or atheists *(another poll sets the proportion of atheists equal to 27%)*, 10% identified as being from other religions or being without opinion, 4% identified as Muslim, 3% identified as Protestant, 1% identified as Buddhist, 1% identified as Jewish.

According to the most recent Eurobarometer Poll 2005, 34% of French citizens responded that "they believe there is a god", whereas 27% answered that "they believe there is some sort of spirit or life force" and 33% that "they do not believe there is any sort of spirit, god, or life force". One other study shows 32% of people in France declaring themselves to be atheists, and another 32% declaring themselves "sceptical about the existence of God but not an atheist".

Estimates of the number of Muslims in France vary widely. According to the 1999 French census returns, there were 3.7 million people of "possible Muslim faith" in France (6.3% of the total population). In 2003, the French Ministry of the Interior estimated the total number of Muslims to be between five and six million (8–10%). The current Jewish community in France numbers around 600,000 according to the World Jewish Congress and is the largest in Europe.

Since 1905 the French government has followed the principle of *laicite*, in which it is prohibited from recognizing any *religion* (except for legacy statutes like that of military chaplains and the local law in Alsace-Moselle). Instead, it merely recognizes *religious organizations*, according to formal legal criteria that do not address religious doctrine. Conversely, religious organizations should refrain from intervening in policy-making.

Certain body of beliefs such as Scientology, Children of God, the Unification Church, or the Order of the Solar Temple are considered cults ("*sectes*" in French), and therefore do not have the same status as religions in France. *Secte* is considered a pejorative term in France.

Public Health

The French healthcare system was ranked first worldwide by the World Health Organization in 1997 and then again in

2000. Care is generally free for people affected by chronic diseases (*Affections de longues durees*) such as cancer, AIDS or Cystic Fibrosis. Average life expectancy at birth is 79.73 years.

As of 2007, there are approximately 140,000 inhabitants (0.4%) of France who are living with HIV/AIDS.

France, as all EU countries, is under an EU directive to reduce sewage discharge to sensitive areas. As of 2006, France is only 40% in compliance with this directive, placing it as one of the lowest achieving countries within the EU with regard to this wastewater treatment standard.

The death of Chantal Sebire revived the debate over euthanasia in France. It was reported on 21 March 2008.

Culture

Architecture

There is, technically speaking, no architecture named *French Architecture*, although that has not always been true. Gothic Architecture's old name was *French Architecture* (or Opus Francigenum). The term "Gothic" appeared later as a stylistic insult and was widely adopted. Northern France is the home of some of the most important Gothic cathedrals and basilicas, the first of these being the Saint Denis Basilica (used as the royal necropolis); other important French Gothic cathedrals are Notre-Dame de Chartres and Notre-Dame d'Amiens. The kings were crowned in another important Gothic church: Notre-Dame de Reims. Aside from churches, Gothic Architecture had been used for many religious palaces, the most important one being the Palais des Papes in Avignon.

During the Middle Ages, fortified castles were built by feudal nobles to mark their powers against their rivals. When King Philip II took Rouen from King John, for example, he demolished the ducal castle to build a bigger one. Fortified cities were also common, unfortunately most French castles did not survive the passage of time. This is why Richard the Lionheart's Chateau-Gaillard was demolished, as well as the Chateau de Lusignan. Some important French castles that survived are Chinon, Chateau d'Angers, the massive Chateau de Vincennes and the so called Cathar castles.

Before the appearance of this architecture France had been using Romanesque architecture like most of Western Europe (with the exception of the Iberian Peninsula, which used Mooresque architecture). Some of the greatest examples of Romanesque churches in France are the Saint Sernin Basilica in Toulouse and the remains of the Cluniac Abbey (largely destroyed during the Revolution and the Napoleonic Wars).

The end of the Hundred Years' War marked an important stage in the evolution of French architecture. It was the time of the French Renaissance and several artists from Italy and Spain were invited to the French court; many residential palaces, Italian-inspired, were built, mainly in the Loire Valley. Such residential castles were the Chateau de Chambord, the Chateau de Chenonceau, or the Chateau d'Amboise. Following the renaissance and the end of the Middle Ages, Baroque Architecture replaced the gothic one. However, in France, baroque architecture found a greater success in the secular domain than in the religious one. In the secular domain the Palace of Versailles has many baroque features. Jules Hardouin Mansart can be said to be the most influential French architect of the baroque style, with his very famous baroque dome of Les Invalides. Some of the most impressive provincial baroque architecture is found in places that were not yet French such as the Place Stanislas in Nancy. On the military architectural side Vauban designed some of the most efficient fortresses of Europe and became a very influential military architect.

After the Revolution the Republicans favoured Neoclassicism although neoclassicism was introduced in France prior to the revolution with such building as the Parisian Pantheon or the Capitole de Toulouse. Built during the French Empire the Arc de Triomphe and Sainte Marie-Madeleine represent this trend the best.

Under Napoleon III a new wave of urbanism and architecture was given birth. If some very extravagant buildings such as the neo-baroque Palais Garnier were built, the urban planning of the time was very organised and rigorous. For example Baron Haussmann rebuilt Paris. The architecture associated to this era is named Second Empire in the English speaking world, the term being taken from the Second French Empire. These times also saw a strong Gothic-Revival trend across Europe, in France the associated architect was Eugene Viollet-le-Duc. In the late 19th century Gustave Eiffel designed many bridges (like the Garabit viaduct) and remains one of the most influential bridge designer of his time, although he is best remembered for the Eiffel Tower.

In the 20th century the Swiss Architect Le Corbusier designed several buildings in France. More recently French architects have combined both modern and old architectural styles. The Louvre Pyramid is a good example of modern architecture added to an older building. Certainly the most difficult buildings to integrate within French cities are skyscrapers, as they are visible from afar. France's largest financial district is La Defense, where a significant number of skyscrapers are located. Other massive buildings that are a challenge to integrate into their environment are large bridges; a good example of the way this has been done is the Millau Viaduct. Some famous modern French architects include Jean Nouvel or Paul Andreu.

Literature

The earliest French literature dates from the Middle Ages when the area that is modern France did not have a single, uniform language. There were several languages and dialects and each writer used his own spelling and grammar. The author of many French mediaeval texts is unknown, for example Tristan and Iseult and Lancelot and the Holy Grail. Much mediaeval French poetry and literature was inspired by the legends of the Matter of France, such as The Song of Roland and the various Chansons de geste. The "Roman de Renart", written in 1175 by Perrout de Saint Cloude tells the story of the mediaeval character Reynard ('the Fox') and is another example of early French writing. The names of some authors from this period are known, for example Chretien de Troyes and Duke William IX of Aquitaine, who wrote in Occitan.

An important 16th century writer was Francois Rabelais who influenced modern French vocabulary and metaphor. During the 17th century Pierre Corneille, Jean Racine and Moliere's plays, Blaise Pascal and René Descartes's moral and philosophical books deeply influenced the aristocracy leaving an important heritage for the authors of the following decades.

Jean de La Fontaine was an important poet from this century. French literature and poetry flourished in the 18th and 19th centuries. The 18th century saw the works of writers, essayists and moralists such as Voltaire, Denis Diderot and Jean-Jacques Rousseau. Charles Perrault was a prolific writer of children's stories such as: "Puss in Boots", "Cinderella", "Sleeping Beauty" and "Bluebeard".

At the turn of the 19th century symbolist poetry was an important movement in French literature, with poets such as Charles Baudelaire, Paul Verlaine and Stephane Mallarme. The 19th century saw the writing of many French novels of world renown with Victor Hugo (*Les Miserables*), Alexandre Dumas (*The Three Musketeers* and *The Count of Monte-Cristo*), and Jules Verne (*Twenty Thousand Leagues Under the Sea*) among the most well-known in France and beyond. Other 19th century fiction writers include Emile Zola, Guy de Maupassant, Theophile Gautier and Stendhal.

The Prix Goncourt is a French literary prize first awarded in 1903. Important writers of the 20th century include Marcel Proust, Louis-Ferdinand Celine, Albert Camus, and Jean-Paul Sartre. Antoine de Saint Exupery wrote *Little Prince* which has remained popular for decades with children and adults around the world.

Sport

Popular sports include football, both codes of rugby football and in certain regions basketball and handball. France has hosted events such as the 1938 and 1998 FIFA World Cups, and hosted the 2007 Rugby Union World Cup. Stade de France in Paris is the largest stadium in France and was the venue for the 1998 FIFA World Cup final, and hosted the 2007 Rugby World Cup final in October 2007. France also hosts the annual Tour de France, the most famous road bicycle race in the world. France is also famous for its 24 Hours of Le Mans sports car endurance race held in the Sarthe department. Several major tennis tournaments take place in France, including the Paris Masters and the French Open, one of the four Grand Slam tournaments.

France has a close association with the Modern Olympic Games; it was a French aristocrat, Baron Pierre de Coubertin, who suggested the Games' revival, at the end of the 19th century. After Athens was awarded the first Games, in reference to the Greek origins of the ancient Olympics, Paris hosted the second Games in 1900. Paris was also the first home of the International Olympic Committee, before it moved to Lausanne. Since that 1900 Games, France has hosted the Olympics on four further occasions: the 1924 Summer Olympics, again in Paris and three Winter Games (1924 in Chamonix, 1968 in Grenoble and 1992 in Albertville).

Both the national football team and the national rugby union team are nicknamed "*Les Bleus*" in reference to the team's shirt colour as well as the national French tricolour flag. The football team is among the most successful in the world, particularly at the turn of the 21st century, with one FIFA World Cup victory in 1998, one FIFA World Cup second place in 2006, and two European Championships in 1984 and 2000. The top national football club competition is the Ligue Rugby is also very popular, particularly in Paris and the southwest of France. The national rugby team has competed at every Rugby World Cup, and takes part in the annual Six Nations Championship. Following from a strong domestic tournament the French rugby team has won sixteen Six Nations Championships, including eight grand slams; and have reached the semi-finals and final of the Rugby World Cup.

Marianne

Marianne is a symbol of the French Republic. She is an allegorical figure of liberty and the Republic and first appeared at the time of the French Revolution. The earliest representations of Marianne are of a woman wearing a Phrygian cap. The origins of the name Marianne are unknown, but Marie-Anne was a very common first name in the 18th century. Anti-revolutionaries of the time derisively called her La Gueuse (the Commoner). It is believed that revolutionaries from the South of France adopted the Phrygian cap as it symbolised liberty, having been worn by freed slaves in both Greece and Rome. Mediterranean seamen and convicts manning the galleys also wore a similar type of cap.

Under the Third Republic, statues, and especially busts, of Marianne began to proliferate, particularly in town halls. She was represented in several different manners, depending on whether the aim was to emphasise her revolutionary nature or her "wisdom". Over time, the Phrygian cap was felt to be too seditious, and was replaced by a diadem or a crown. In recent times, famous French women have been used as the model for those busts. Recent ones include Sophie Marceau, and Laetitia Casta. She also features on everyday articles such as postage stamps and coins.

International Rankings

- Institute for Economics and Peace-Global Peace Index: Rank 30 out of 144 countries
- Total GDP, 2007: 5th (out of 179) (IMF data)
- Total value of foreign trade (imports and exports), 2002: 4th (out of 185)
- Reporters Without Borders worldwide press freedom index 2005: Rank 35 out of 167 countries
- Transparency International Corruption Perceptions Index 2006 – 18th of 163 countries.

6

Grand-Duchy of Luxembourg

Luxembourg, officially the Grand Duchy of Luxembourg, is a landlocked country in western Europe, bordered by Belgium, France, and Germany. Luxembourg has a population of over half a million people in an area of approximately 2,586 square kilometres (999 sq mi).

Luxembourg is a parliamentary representative democracy with a constitutional monarch; it is ruled by a Grand Duke. It is the world's only remaining sovereign Grand Duchy. The country has a highly developed economy, with the highest Gross Domestic Product per capita in the world as per IMF and WB. Its historic and strategic importance dates back to its founding as a Roman era fortress site and Frankish count's castle site in the Early Middle Ages. It was an important bastion along the Spanish Road when Spain was the principal European power influencing the whole western hemisphere and beyond in the 16th–17th centuries.

Luxembourg is a founding member of the European Union, NATO, OECD, the United Nations, Benelux, and the Western European Union, reflecting the political consensus in favour of economic, political, and military integration. The city of Luxembourg, the capital and largest city, is the seat of several institutions and agencies of the European Union.

Luxembourg lies on the cultural divide between Romance Europe and Germanic Europe, borrowing customs from each of the distinct traditions. Luxembourg is a trilingual country; German, French and Luxembourgish are official languages.

Although a secular state, Luxembourg is predominantly Roman Catholic.

History

The recorded history of Luxembourg begins with the acquisition of Lucilinburhuc (today Luxembourg Castle) by Siegfried, Count of Ardennes in 963. Around this fort, a town gradually developed, which became the centre of a small state of great strategic value. In the 14th and early 15th centuries three members of the House of Luxembourg reigned as Holy Roman Emperors. In 1437, the House of Luxembourg suffered a succession crisis, precipitated by the lack of a male heir to assume the throne, which led to the territory being sold to Philip the Good of Burgundy.

In the following centuries, Luxembourg's fortress was steadily enlarged and strengthened by its successive occupants, the Bourbons, Habsburgs, Hohenzollerns, and the French, among others. After the defeat of Napoleon in 1815, Luxembourg was disputed between Prussia and the Netherlands. The Congress of Vienna formed Luxembourg as a Grand Duchy in personal union with the Netherlands. Luxembourg also became a member of the German Confederation, with a Confederate fortress manned by Prussian troops.

The Belgian Revolution of 1830–1839 reduced Luxembourg's territory by more than half, as the predominantly francophone western part of the country was transferred to Belgium. Luxembourg's independence was reaffirmed by the 1839 First Treaty of London. In the same year, Luxembourg joined the Zollverein. Luxembourg's independence and neutrality were again affirmed by the 1867 Second Treaty of London, after the Luxembourg Crisis nearly led to war between Prussia and France. After the latter conflict, the Confederate fortress was dismantled.

The King of the Netherlands remained Head of State as Grand Duke of Luxembourg, maintaining personal union between the two countries until 1890. At the death of William III, the Dutch throne passed to his daughter Wilhelmina, while Luxembourg (at that time restricted to male heirs by the Nassau Family Pact) passed to Adolph of Nassau-Weilburg.

Luxembourg was invaded and occupied by Germany during the First World War, but was allowed to maintain its independence and political mechanisms. It was again invaded and subject to German occupation in the Second World War in 1940, and was formally annexed into the Third Reich in 1942.

During World War II, Luxembourg abandoned its policy of neutrality, when it joined the Allies in fighting Germany. Its government, exiled to London, set up a small group of volunteers who participated in the Normandy invasion. It became a founding member of the United Nations in 1946, and of NATO in 1949.

In 1957, Luxembourg became one of the six founding countries of the European Economic Community (later the European Union), and, in 1999, it joined the euro currency area. In 2005, a referendum on the EU treaty establishing a constitution for Europe was held in Luxembourg.

Government and Politics

Luxembourg is a parliamentary democracy headed by a constitutional monarch. Under the constitution of 1868, executive power is exercised by the Grand Duke and the cabinet, which consists of several other ministers.

The Governor has the power to dissolve the legislature and reinstate a new one, as long as the Grand Duke has judicial approval. However, since 1919, sovereignty has resided with the Supreme Court.

Legislative power is vested in the Chamber of Deputies, a unicameral legislature of sixty members, who are directly elected to five-year terms from four constituencies. A second body, the Council of State (*Conseil d'État*), composed of twenty-one ordinary citizens appointed by the Grand Duke, advises the Chamber of Deputies in the drafting of legislation.

The Grand Duchy has three lower tribunals (*justices de paix*; in Esch-sur-Alzette, the city of Luxembourg, and Diekirch), two district tribunals (Luxembourg and Diekirch) and a Superior Court of Justice (Luxembourg), which includes the Court of Appeal and the Court of Cassation. There is also an Administrative Tribunal and an Administrative Court, as well as a Constitutional Court, all of which are located in the capital.

Districts, Cantons and Communes

Luxembourg is divided into 3 districts, which are further divided into 12 cantons and then 116 communes. Twelve of the communes have city status, of which the city of Luxembourg is the largest.

Military

The contribution Luxembourg makes to its defence and to NATO consists of a small army (currently consisting of around 800 people). As a landlocked country, it has no navy. Luxembourg also lacks an air force, though the seventeen NATO AWACS aeroplanes are for convenience registered as aircraft of Luxembourg. In accordance with a joint agreement with Belgium, both countries have put forth funding for one A400M military cargo plane, currently on order. Luxembourg still jointly maintains three NATO Boeing 707 model TCAs (for cargo and training purposes) based at NATO Air Base Geilenkirchen.

Geography and Climate

Luxembourg is a small country located in the Low Countries, part of North-West Europe It borders Belgium (148 kilometres) to the west and north, France (73 km) to the south, and Germany (138 km) to the east. Luxembourg is landlocked, separated from the North Sea by Belgium. The topography of the country is divided very clearly between the hilly Oesling of the northern third of the Grand Duchy and the flat Gutland, which occupies the southern two-thirds.

The country's longest river is the Sauer, which is a tributary of the Moselle, the basin of which includes almost all of Luxembourg's area. Other major rivers include the Alzette in the south and the Wiltz in the north. The capital, and by far the largest city, is Luxembourg City, which is located in the Gutland, as are most of the country's main population centres, including Esch-sur-Alzette, Dudelange, and Differdange. Besides Luxembourg City, the other main towns are primarily located in the southern Red Lands region, which lines the border between Luxembourg and France to the south.

Physical Geography

Topography

Despite its small size, Luxembourg has a varied topography, with two main features to its landscape. The northern section of the country is formed by part of the plateau of the Ardennes, where the mountain heights range from 1,500 to almost 2,000 feet (460–610 m) high. The rest of the country is made up of undulating countryside with broad valleys. The capital, Luxembourg City, is located in the southern part of the country.

Green Heart of Europe

The most prominent landmark, the high plateau of the Ardennes in the north, took millions of years to carve. At its highest point, it reaches a height of 1,834 feet (559 m). Commonly known as the Oesling, the Ardennes region covers 325 sq. miles (842 km^2), about 32% of the entire country. Rugged scenery predominates because river erosion over thousands of years has left a varied, low mountain landscape, densely covered with vegetation, sometimes with considerable variations in height. These differences in relief, together with stretches of water interspersed with forests, fields, and pastures are the main features that make the landscape so distinctive. Typical of this high area, however, is the infertile soil and poor drainage resulting in numerous peat bogs, which were once exploited as fuel. These factors, combined with heavy rainfall and frost, made this an inhospitable environment for the first settlers.

Even today, the living conditions in such an environment are not particularly favourable. Nevertheless, some 7,800 people make a living of the land through either forestry, small-scale farming, or environment work. Because the soil is so difficult to cultivate, most of the land is used for cattle pasture. The Ardennes region also includes the Upper Sûre National Park, an important conservation area and a hiker's retreat.

South of the Sûre River, the country is known as the Gutland. The region covers slightly over two-thirds of the country.

The terrain gently rises and falls with an average height of 700 feet (200 m). Agriculture is the main activity as term Gutland arises from the fertile soil and warm, dry summers experienced is this part of the Duchy compared to the Oesling region.

As a result, vegetables and fruit, such as strawberries, apples, plums, and cherries, are grown in large quantities. River erosion in this area has created deep gorges and caves, resulting in some spectacular scenery.

In the extreme south of the country lies "the land of the red rocks"-a reference to the deposits of minerals found here. Rich in iron ore, the district has been a mining and heavy industrial region since Roman if not earlier times and stretches for over 12 miles (20 km). The tall chimneys of the iron and steel works are typical landmarks of the industrial south. To the east lies the grape-growing valley of the Moselle. Numerous villages nestle in the deep valleys and behind the vineyards along the river banks. Most villages have at least one winery. Also in the east is the "Little Switzerland" area, characterized by wooded glens and ravines in unusual rock formations.

Rivers and Lakes

Luxembourg has a number of minor rivers, such as the Eisch, the Alzette, and the Petrusse, but the main river is the Moselle with its tributaries-the Sûre and the Our. Together, their courses serve as a natural boundary between Luxembourg and Germany. Along their banks, many of the country's medieval castles can be found.

The Moselle River actually rises in northeast France and flows north through Luxembourg for 19 miles (31 km) to join the mighty Rhine at Koblenz, Germany. The Moselle is 320 miles (515 km) long, and is navigable, due to canalization for 40 miles (64 km). Green slopes, covered with vines, flank the meandering course of the river.

Rising in Belgium, the Sûre River flows for 107 miles (172 km) in an easterly direction through Luxembourg and into the Moselle. Its sinuous course essentially cuts Luxembourg from east to west. The Our River, flowing along the northeastern border, is a tributary of the Sûre. Its valley is surrounded by unspoiled countryside.

The Upper Sûre lake is the largest stretch of water in the Grand Duchy. Surrounded by luxuriant vegetation and peaceful creeks, the lake is a centre for water sports, such as sailing, canoeing, and kayaking. Such outdoor activities, which has made it an attractive spot for tourists, have led to the growth of a local crafts industry.

The town of Esch-sur-Sûre nestles at one end of the lake. Immediately above it, the river has been dammed to form a hydroelectric reservoir extending some 6 miles (10 km) up the valley. The Upper Sûre dam was built in the 1960s to meet the country's drinking water requirements.

Luxembourg is one of the smallest countries in Europe, and ranked 175th in size of all the 194 independent countries of the world; the country is about 2,586 square kilometres (998 sq mi) in size, and measures 82 km (51 miles) long and 57 km (35 miles) wide. To the east, Luxembourg borders the German *Bundesländer* of Rhineland-Palatinate and Saarland, and, to the south, it borders the French *region* of Lorraine. The Grand Duchy borders the Belgian Walloon Region, in particular the latter's provinces of Luxembourg and Liege, more in particular the German-speaking Community of Belgium, to the west and to the north respectively.

The northern third of the country is known as the 'Oesling', and forms part of the Ardennes. It is dominated by hills and low mountains, including the Kneiff near Wilwerdange, which is the highest point, at 560 metres (1,837 ft). Other mountains are the 'Buurgplaaz' at 559 metres near Huldange and the 'Napoleonsgaard' at 548 metres near Rambrouch. The region is sparsely populated, with only one town (Wiltz) with a population of more than four thousand people.

The southern two-thirds of the country is called the "Gutland", and is more densely populated than the Oesling. It is also more diverse, and can be divided into five geographic sub-regions. The Luxembourg plateau, in south-central Luxembourg, is a large, flat, sandstone formation, and the site of the city of Luxembourg. Little Switzerland, in the east of Luxembourg, has craggy terrain and thick forests. The Moselle valley is the lowest-lying region, running along the south-eastern border. The Red Lands, in the far south and southwest, are Luxembourg's industrial heartland and home to many of Luxembourg's largest towns.

The border between Luxembourg and Germany is formed by three rivers: the Moselle, the Sauer, and the Our. Other major rivers are the Alzette, the Attert, the Clerve, and the Wiltz. The valleys of the mid-Sauer and Attert form the border between the Gutland and the Oesling.

Luxembourg has a marine west coast climate, marked by high precipitation, particularly in late summer.

Demographics

Ethnicity

The people of Luxembourg are called Luxembourgers. The native population has a Celtic base with a French and Germanic blend. The immigrant population increased in the twentieth century due to the arrival of immigrants from Belgium, France, Germany, Italy, and Portugal, with the majority coming from this last country. In 2001 census, there were 58,657 inhabitants with Portuguese nationality.

Since the beginning of the Yugoslav wars, Luxembourg has seen many immigrants from Bosnia and Herzegovina, Montenegro, and Serbia. Annually, over 10,000 new immigrants arrive in Luxembourg, mostly from the EU states, as well and Eastern Europe. As of 2000, there were 162,000 immigrants in Luxembourg, accounting for 37% of the total population. There were an estimated 5,000 undocumented immigrants, including asylum seekers, in Luxembourg as of 1999.

Language

Three languages are recognised as official in Luxembourg: French, German, and Luxembourgish, a Franconian language of the Moselle region very similar to the local German dialect spoken in the neighbouring part of Germany. More than 5,000 words in the Luxembourgish have their origin in the French language. The first printed sentences in Luxembourgisch were released by the weekly journal the 'Luxemburger Wochenblatt' in 1821.

Apart from being one of the three official languages, Luxembourgish is also considered the national language of the Grand Duchy; it is the mother tongue or "language of the heart" for nearly all Luxembourgers.

Each of the three languages is used as the primary language in certain spheres. Luxembourgish is the language that Luxembourgers generally use to speak to each other, but it is not often written down. Most official (written) business is carried out in French. German is usually the first language taught in school and is the language of much of the media and of the church.

Luxembourg's education system is trilingual: the first years of primary school are in Luxembourgish, before changing to German, while in secondary school, the language of instruction changes to French. However, as proficiency in all three languages is required for graduation from secondary school, half the students leave school without a certified qualification, with the children of immigrants being particularly disadvantaged.

In addition to the three official languages, English is taught in the compulsory schooling and much of the population of Luxembourg can speak English, at any rate in Luxembourg City. Portuguese and Italian, the languages of the two largest immigrant communities, are also spoken by large parts of the population, but by relatively few from outside their respective communities.

Religion

Luxembourg is a secular state, but the state recognises certain religions as officially mandated religions. This gives the state a hand in religious administration and appointment of clergy, in exchange for which the state pays certain running costs and wages. Currently, religions covered by such arrangements are Roman Catholicism, Judaism, Greek Orthodoxy, Anglicanism, Russian Orthodoxy, Protestantism and Islam.

Since 1980 it has been illegal for the government to collect statistics on religious beliefs or practices. An outdated estimation by the CIA Factbook for the year 2000 is that 87% of Luxembourgers are Catholics, including the royal family, the remaining 13% being made up of Protestants, Orthodox Christians, Jews, Muslims and those of other or no religion.

According to the most recent Eurobarometer Poll 2005, 44% of Luxembourg citizens responded that "they believe there is a God", whereas 28% answered that "they believe there is some sort of spirit or life force" and 22% that "they do not believe there is any sort of spirit, god, or life force".

Economy

Luxembourg's stable, high-income economy features moderate growth, low inflation, and low unemployment. The industrial sector, which was dominated until the 1960s by steel, has diversified to include chemicals, rubber, and other products. During the past decades, growth in the financial sector has more than compensated for the decline in steel. Services, especially banking and other financial exports, account for the majority of economic output. In March 2010, the Sunday Telegraph reported that most of Kim Jong-Il's $4bn in secret accounts is in Luxemburg banks. Luxembourg is the world's second largest investment fund center (after the USA), the most important private banking center in the Eurozone and Europe's leading center for reinsurance companies. Moreover, the Luxembourgish government has tried to attract internet start-ups. Skype and eBay are two of the many internet companies that have shifted their local or global headquarters to Luxembourg.

Agriculture is based on small, family-owned farms. Luxembourg has especially close trade and financial ties to Belgium and the Netherlands, and as a member of the EU it enjoys the advantages of the open European market. Luxembourg possesses the highest GDP per capita in the world (US$87,995 as of 2006), the eighteenth highest Human Development Index, and the fourth highest rated in the quality of life index. As of March 2006, unemployment is 4.8% of the labour force. For the fiscal year of 2005 and 2006, Luxembourg has run a budget deficit for the first time in many years, mostly because of slower international economic growth.

Transport

Luxembourg has efficient road, rail and air transport facilities and services. The road network has been significantly modernized in recent years with 147 km of motorways connecting the capital to adjacent countries. The advent of the high-speed TGV link to Paris has led to renovation of the city's railway station while a new passenger terminal at Luxembourg Airport has recently been opened. There are plans to introduce trams in the capital and light-rail lines in adjacent areas within the next few years.

Telecommunication and Internet

The telecommunications industry in Luxembourg is liberalized and the electronic communications networks are significantly developed. A healthy competition between the different operators is guaranteed by the legislative framework Paquet Telecom of the Government of 2005 which transposes the European Telecom Directives into Luxembourgean law. This encourages the investment in networks and services. The regulator ILR-Institut Luxembourgeois de Régulation ensures the compliance to these legal rules

Luxembourg has modern and widely deployed optical fibre and cable networks throughout the country. With its central position in Europe, stable economy and low taxes, Luxembourg is an ideal gateway to the European telecommunication markets.

It ranks 2nd in the world in the development of the Information and Communication Technologies in the ITU ICT Development Index and 8th in the Global Broadband Quality Study 2009 by the University of Oxford and the University of Oviedo.

Luxembourg is connected to all major European Internet Exchanges (AMS-IX Amsterdam, DE-CIX Frankfurt, LINX London), datacenters and POPs through redundant optical networks. In addition, the country is connected to the virtual meetme room services (vmmr) of the international data hub operator Ancotel. This enables Luxembourg to interconnect with all major telecommunication operators and data carriers worldwide. The interconnection points are in Frankfurt, London, New York and Hong Kong.

Several providers interconnect Luxembourg to the major European data hubs;

- TERALINK (P&T Luxembourg, also called EPT Luxembourg: incumbent operator)
- LuxConnect (shareholder : Government)
- Artelis/Cegecom (alternative telecommunications provider in Luxembourg and Saarland)
- Satellite connectivity-Teleports (SES Astra, Broadcasting Center Europe and P&T Luxembourg-Uplink Cloche d'Or).

More than 10 data center operators are established in Luxembourg.

Internet Exchange Points

Luxembourg is connected through Teralink to several Tier 1 upstream providers like Level 3 and Global Crossing. The Internet IPV 6 protocol has been introduced to the country by Restena and P&T Luxembourg.

Luxembourg has two Internet exchange points.

- LIX is a neutral Internet Exchange operated by the RESTENA Foundation. Its aim is to easily interconnect the different Internet Service Providers (ISPs) within Luxembourg.
- LUX-CIX is Luxembourg's neutral and commercial Internet Exchange Point which was founded in 2009 by Cegecom, Datacenter Luxembourg, Global Media Systems, INEXIO, LuxConnect, P&T Luxemboug and Root eSolutions. It offers a short, fast and efficient route to the major European Internet networks.

Information Society

The online portal De Guichet of the Grand Duchy of Luxembourg is a single one-stop online shop for citizens and companies to realize different sorts of administrative steps (procedures, online forms, downloadable forms and advices) by Internet.

Telematic Services

PSA Peugeot Citroën, with P&T Luxembourg as its partner, has introduced an integrated mobile telecommunication solution for the development of its telematic services in Europe.

Culture

Luxembourg has been overshadowed by the culture of its neighbours, although, having been for much of its history a profoundly rural country, it retains a number of folk traditions. There are several notable museums, mostly located in the capital; these include the National Museum of History and Art (MNHA), the History Museum of the City of Luxembourg, and the new Grand Duke Jean Museum of Modern Art (Mudam). The National Museum of Military History (MNHM) in Diekirch is especially known for its representations of the Battle of the Bulge. The city of Luxembourg itself is on the UNESCO World Heritage List, on account of the historical importance of its fortifications.

The country has produced some internationally known artists, including the painters Joseph Kutter and Michel Majerus, as well as the photographer Edward Steichen. Steichen's *The Family of Man* exhibition is now permanently housed in Clervaux, and it has been placed on UNESCO's Memory of the World register. Movie star Loretta Young was of Luxembourgian descent.

Luxembourg was the first city to be named European Capital of Culture twice. The first time was in 1995. In 2007, the European Capital of Culture was to be a cross-border area consisting of the Grand Duchy of Luxembourg, the Rheinland-Pfalz and Saarland in Germany, the Walloon Region and the German-speaking part of Belgium, and the Lorraine area in France.

The event was an attempt to promote mobility and the exchange of ideas, crossing borders in all areas, physical, psychological, artistic and emotional.

World Expo

Luxembourg will be represented at the World Expo 2010 in Shanghai, China from May 1 to October 31, 2010 with its own pavilion. The pavilion is based on the translation of the word Luxembourg into Chinese, which means 'Lu Sen Bao': Forest and Fortress. It represents Luxembourg as the 'Green Heart in Europe'.

Media

Luxembourg is well known in Europe for its radio and television stations (Radio Luxembourg, RTL Group). It is also the uplink home of SES Astra, carrier of major European satellite services for Germany and Britain.

Due to the audiovisual law of 1988, which established a special tax regime for audiovisual investment, the film and co-production in Luxembourg has grown steadily. There exists approximately 30 by agreed production companies in Luxembourg.

Taxes on Alcohol, Cigarettes and Petrol

Luxembourg consumes the most alcohol in Europe per capita. However, the large proportion of alcohol purchase by customers from neighboring countries contributes to the statistically high level of alcohol consumption per capita; this level of consumption is thus not representative of the actual alcohol consumption of the Luxembourg population.

7

Germany

Germany, officially the Federal Republic of Germany is a country in Central Europe. It is bordered to the north by the North Sea, Denmark, and the Baltic Sea; to the east by Poland and the Czech Republic; to the south by Austria and Switzerland; and to the west by France, Luxembourg, Belgium, and the Netherlands. The territory of Germany covers 357,021 square kilometers (137,847 sq mi) and is influenced by a temperate seasonal climate. With 81.8 million inhabitants in January 2010, it has the largest population among member states of the European Union, and it is also home to the third-largest number of international migrants worldwide.

A region named Germania, inhabited by several Germanic peoples, has been known and documented before AD 100. Beginning in the 10th century, German territories formed a central part of the Holy Roman Empire, which lasted until 1806. During the 16th century, northern Germany became the centre of the Protestant Reformation. As a modern nation-state, the country was first unified amidst the Franco-Prussian War in 1871. In 1949, after World War II, Germany was divided into two separate states—East Germany and West Germany—along the lines of Allied occupation. Germany was reunified in 1990. West Germany was a founding member of the European Community (EC) in 1957, which became the European Union in 1993. It is part of the Schengen zone and adopted the European currency, the euro, in 1999.

Germany is a federal parliamentary republic of sixteen states (*Bundesländer*). The capital and largest city is Berlin.

Germany is a member of the United Nations, NATO, G8, G20, OECD, and the WTO. It is a major power with the world's fourth largest economy by nominal GDP and the fifth largest in purchasing power parity. It is the second largest exporter and third largest importer of goods. In absolute terms, Germany allocates the second biggest annual budget of development aid in the world, while its military expenditure ranked seventh. The country has developed a high standard of living and established a comprehensive system of social security. It holds a key position in European affairs and maintains a multitude of close partnerships on a global level. Germany is recognised as a scientific and technological leader in several fields.

History

The English word "Germany" derives from the Latin word Germania. The name "Germania" came into use after Julius Caesar adopted it from a Gallic term for the peoples east of the Rhine that probably meant "neighbour".

Germanic Tribes

The ethnogenesis of the Germanic tribes is assumed to have occurred during the Nordic Bronze Age, or at the latest, during the Pre-Roman Iron Age. From southern Scandinavia and northern Germany, the tribes began expanding south, east and west in the 1st century BC, coming into contact with the Celtic tribes of Gaul as well as Iranian, Baltic, and Slavic tribes in Eastern Europe. Little is known about early Germanic history, except through their recorded interactions with the Roman Empire, etymological research and archaeological finds.

Under Augustus, the Roman General Publius Quinctilius Varus began to invade Germania (a term used by the Romans to define a territory running roughly from the Rhine to the Ural Mountains), and it was in this period that the Germanic tribes became familiar with Roman tactics of warfare while maintaining their tribal identity. In AD 9, three Roman legions led by Varus were defeated by the Cheruscan leader Arminius in the Battle of the Teutoburg Forest. Modern Germany, as far as the Rhine and the Danube, thus remained outside the Roman Empire. By AD 100, the time of Tacitus' *Germania*, Germanic tribes settled along the Rhine and the Danube (the Limes Germanicus), occupying most of the area of modern Germany;

Austria, southern Bavaria and the western Rhineland, however, were Roman provinces. The 3rd century saw the emergence of a number of large West Germanic tribes: Alamanni, Franks, Chatti, Saxons, Frisians, Sicambri, and Thuringii. Around 260, the Germanic peoples broke through the Limes and the Danube frontier into Roman-controlled lands.

Holy Roman Empire (962–1806)

On 25 December 800, Charlemagne founded the Carolingian Empire, which was divided in 843. The medieval empire resulted from the eastern portion of this division and existed in varying forms from 962 until 1806. Its territory stretched from the Eider River in the north to the Mediterranean coast in the south. Often referred to as the Holy Roman Empire (or the Old Empire), it was officially called the *Sacrum Romanum Imperium Nationis Germanicæ* (Holy Roman Empire of the German Nation) starting in 1448, to adjust the title to its then reduced territory.

Under the reign of the Ottonian emperors (919–1024), the duchies of Lorraine, Saxony, Franconia, Swabia, Thuringia, and Bavaria were consolidated, and the German king was crowned Holy Roman Emperor of these regions in 962. Under the reign of the Salian emperors (1024–1125), the Holy Roman Empire absorbed northern Italy and Burgundy, although the emperors lost power through the Investiture Controversy. Under the Hohenstaufen emperors (1138–1254), the German princes increased their influence further south and east into territories inhabited by Slavs, preceding German settlement in these areas and further east *(Ostsiedlung)*. Northern German towns grew prosperous as members of the Hanseatic League. Starting with the Great Famine in 1315, then the Black Death of 1348–50, the population of Germany plummeted.

The edict of the Golden Bull in 1356 provided the basic constitution of the empire that lasted until its dissolution. It codified the election of the emperor by seven prince-electors who ruled some of the most powerful principalities and archbishoprics. Beginning in the 15th century, the emperors were elected nearly exclusively from the Habsburg dynasty of Austria. The monk Martin Luther publicised his 95 Theses in 1517, challenging practices of the Roman Catholic Church, initiating the Protestant Reformation. A separate Lutheran church became the official religion in many German states after 1530. Religious conflict led to the Thirty Years' War (1618–1648), which devastated German lands. The population of the German states was reduced by about 30%.

The Peace of Westphalia (1648) ended religious warfare among the German states, but the empire was *de facto* divided into numerous independent principalities. From 1740 onwards, the dualism between the Austrian Habsburg Monarchy and the Kingdom of Prussia dominated German history. In 1806, the *Imperium* was overrun and dissolved as a result of the Napoleonic Wars.

Restoration and Revolution (1814–1871)

Following the fall of Napoleon Bonaparte, the Congress of Vienna convened in 1814 and founded the German Confederation (Deutscher Bund), a loose league of 39 sovereign states. Disagreement with restoration politics partly led to the rise of liberal movements, demanding unity and freedom.

These, however, were followed by new measures of repression on the part of the Austrian statesman Metternich. The *Zollverein*, a tariff union, profoundly furthered economic unity in the German states. During this era many Germans had been stirred by the ideals of the French Revolution, and nationalism became a more significant force, especially among young intellectuals. For the first time, the colours of black, red and gold were chosen to represent the movement, which later became the national colours.

In light of a series of revolutionary movements in Europe, which successfully established a republic in France, intellectuals and commoners started the Revolutions of 1848 in the German states. The monarchs initially yielded to the revolutionaries' liberal demands. King Frederick William IV of Prussia was offered the title of Emperor, but with a loss of power; he rejected the crown and the proposed constitution, leading to a temporary setback for the movement. Conflict between King William I of Prussia and the increasingly liberal parliament erupted over military reforms in 1862, and the king appointed Otto von Bismarck the new Prime Minister of Prussia. Bismarck successfully waged war on Denmark in 1864. Prussian victory in the Austro-Prussian War of 1866 enabled him to create the

North German Federation (Norddeutscher Bund) and to exclude Austria, formerly the leading German state, from the affairs of the remaining German states.

German Empire (1871–1918)

The state known as *Germany* was unified as a modern nation-state in 1871, when the German Empire was forged, with the Kingdom of Prussia as its largest constituent. After the French defeat in the Franco-Prussian War, the German Empire was proclaimed in Versailles on 18 January 1871. The Hohenzollern dynasty of Prussia ruled the new empire, whose capital was Berlin. The empire was a unification of all the scattered parts of Germany except Austria (*Kleindeutschland*, or "Lesser Germany"). But internally the official political unification came rather sequentially: Germany had no national flag until 1892 and no national hymn until after WW I. Beginning in 1884, Germany began establishing several colonies outside of Europe.

In the *Grunderzeit* period following the unification of Germany, Emperor William I's foreign policy secured Germany's position as a great nation by forging alliances, isolating France by diplomatic means, and avoiding war. Under William II, however, Germany, like other European powers, took an imperialistic course leading to friction with neighbouring countries. Most alliances in which Germany had been previously involved were not renewed, and new alliances excluded the country. Specifically, France established new relationships by signing the Entente Cordiale with the United Kingdom and securing ties with the Russian Empire. Aside from its contacts with Austria-Hungary, Germany became increasingly isolated.

Germany's imperialism reached outside of its own country and joined many other powers in Europe in claiming their share of Africa. The Berlin Conference divided Africa between the European powers. Germany owned several pieces of land in Africa including German East Africa, South-West Africa, Togo, and Cameroon. The Scramble for Africa caused tension between the great powers that may have contributed to the conditions that led to World War I.

The assassination of Austria's crown prince on 28 June 1914 triggered World War I. Germany, as part of the unsuccessful Central Powers, suffered defeat against the Allied Powers in one of the bloodiest conflicts of all time. An estimated two million German soldiers died in World War I. The German Revolution broke out in November 1918, and Emperor William II and all German ruling princes abdicated. An armistice putting an end to the war was signed on 11 November and Germany was forced to sign the Treaty of Versailles in June 1919. Its negotiation, contrary to traditional post-war diplomacy, excluded the defeated Central Powers. The treaty was perceived in Germany as a humiliating continuation of the war by other means and its harshness is often cited as having facilitated the later rise of Nazism in the country.

Weimar Republic (1919–1933)

At the beginning of the German Revolution, Germany was declared a republic and the monarchy collapsed. However, the struggle for power continued, with radical-left communists seizing power in Bavaria, but failing to take control of all of Germany. The revolution came to an end in August 1919, when the Weimar Republic was formally established. The Weimar Constitution came into effect with its signing by President Friedrich Ebert on 11 August 1919.

Suffering from the Great Depression, the harsh peace conditions dictated by the Treaty of Versailles, and a long succession of more or less unstable governments, the people of Germany increasingly lacked identification with their political system and the "Establishment Parties" in their parliamentary democracy. This was exacerbated by a widespread right-wing (monarchist, *volkisch*, and Nazi) *Dolchstoblegende*, which promoted the view that Germany had lost World War I because of the efforts and influence of those who wanted to overthrow the government. The top brass of the Weimar government was accused of betraying the German Nation by signing the Versailles Treaty, while the radical left-wing communists, such as the Spartacist League, had wanted a revolution to abolish "capitalist rule" in favour of a *Raterepublik*, and were also targeted.

Nevertheless, discontentment with the new Weimar government helped fuel the growth of the German Communist Party. Many conservatives were drawn towards the reactionary/revolutionary right, particularly the National Socialist German

Workers Party—the Nazi Party. By 1932, these two parties controlled the majority of parliament (296 total parliamentary seats by July 1932). After a series of unsuccessful cabinets, President Paul von Hindenburg made a crucial decision: on 30 January 1933, seeing little alternative and pushed by right-wing advisors, von Hindenburg appointed Adolf Hitler as Chancellor of Germany, honoring Hitler's request.

Third Reich (1933–1945)

On 27 February 1933, the Reichstag building went up in flames, and a consequent emergency decree abrogated basic citizen rights. An Enabling Act passed in parliament gave Hitler unrestricted legislative power. Only the Social Democratic Party voted against it, while Communist MPs had already been imprisoned. Using his powers to crush any actual or potential resistance, Hitler established a centralised totalitarian state within months. Industry was revitalised with a focus on military rearmament. In 1935, Germany required control of the Saar and in 1936 military control of the Rhineland, both of which had been lost by the Treaty of Versailles.

Leading to World War II and roughly in parallel with military rearmament, German foreign policy became more aggressive and expansionistic. In 1938 and 1939, Austria and Czechoslovakia were brought under control and the invasion of Poland prepared (Hitler-Stalin pact, Operation Himmler). On 1 September 1939, the German Wehrmacht launched a blitzkrieg on Poland, which was swiftly occupied by Germany and by the Soviet Red Army. The UK and France declared war on Germany marking the beginning of World War II in Europe. As the war progressed, Germany and its allies quickly gained control of much of continental Europe.

On 22 June 1941, Germany broke the Hitler-Stalin pact and invaded the Soviet Union. The same year, Japan attacked the American base at Pearl Harbor, and Germany declared war on the United States as a consequence of its alliance with Japan. Although the German army advanced into the Soviet Union quite rapidly, the Battle of Stalingrad marked a major turning point in the war. Subsequently, the German army started to retreat on the Eastern front. In September 1943, Germany's ally Italy surrendered, and German forces were forced to defend an additional front in Italy. D-Day marked another major turning point in the war, opening up a Western front; the Allied forces landed on the beaches of Normandy and made advances towards German territory. Germany's defeat soon followed. On 8 May 1945, the German armed forces surrendered after the Red Army occupied Berlin. Approximately seven million German soldiers and civilians—including ethnic Germans from Eastern Europe—died during World War II.

In what later became known as The Holocaust, the Third Reich regime enacted governmental policies directly subjugating many dissidents and minorities. About seventeen million people were murdered by the Nazis during the Holocaust, including six million Jews and a sizable number of Gypsies, Poles and other Slavs, including Soviet POWs, the mentally ill, homosexuals, and members of the political opposition. World War II and the Nazi genocide were responsible for more than 40 million dead in Europe. The Nuremberg trials of Nazi war criminals were held after World War II.

Division and Reunification (1945–1990)

The war resulted in the death of over five million German soldiers and civilians; large territorial losses; the expulsion of about 15 million Germans from the eastern areas of Germany and other countries; mass rape of German women; and the destruction of multiple major cities. The remaining national territory and Berlin were partitioned by the Allies into four military occupation zones.

The western sectors, controlled by France, the United Kingdom, and the United States, were merged on 23 May 1949, to form the *Federal Republic of Germany* (*Bundesrepublik Deutschland*); on 7 October 1949, the Soviet Zone became the *German Democratic Republic* (*Deutsche Demokratische Republik*, or DDR). They were, mainly outside Germany, informally known as "West Germany" and "East Germany", and the two parts of Berlin as "West Berlin" and "East Berlin". East Germany selected East Berlin as its capital, while West Germany chose Bonn. However, West Germany declared the status of its capital Bonn as provisional, in order to emphasise its stance that the two-state solution was an artificial *status quo* that was to be overcome one day.

West Germany, established as a federal parliamentary republic with a "social market economy", was allied with the United States, the UK and France. The country came to enjoy prolonged economic growth beginning in the early 1950s (*Wirtschaftswunder*). West Germany joined NATO in 1955 and was a founding member of the European Economic Community in 1957. On 1 January 1957, Saarland gave in its adhesion to West Germany by virtue of article 23 *Grundgesetz*.

East Germany was an Eastern bloc state under political and military control by the USSR via the latter's occupation forces and the Warsaw Pact. While claiming to be a democracy, political power was solely executed by leading members (*Politburo*) of the communist-controlled SED (*Socialist Unity Party of Germany*). Their power was ensured by the Stasi, a secret service of immense size, and a variety of SED suborganizations controlling every aspect of society. In return, the basic needs of the population were satisfied at low cost by the state. A Soviet-style command economy was set up; later, the GDR became a Comecon state. While East German propaganda was based on the benefits of the GDR's social programs and the alleged constant threat of a West German invasion, many of her citizens looked to the West for political freedoms and economic prosperity. The Berlin Wall, built in 1961 to stop East Germans from escaping to West Germany, became a symbol of the Cold War.

Tensions between East and West Germany were somewhat reduced in the early 1970s by Chancellor Willy Brandt's *Ostpolitik*, which included the *de facto* acceptance of Germany's territorial losses in World War II.

In the summer of 1989, Hungary decided (May 2) to dismantle the Iron Curtain and open the borders (August 23), causing an exodus of thousands of East Germans (September 11) going to West Germany via Hungary. The effects of the Hungarian events had devastating effects on the GDR, with mass demonstrations. The East German authorities unexpectedly eased the border restrictions in November, allowing East German citizens to travel to the West. Originally intended as a pressure valve to retain East Germany as a state, the opening of the border actually led to an acceleration of the *Wende* reform process in East Germany, which finally concluded with the *Two Plus Four Treaty* a year later on 12 September 1990, under which the four occupying powers renounced their rights under the Instrument of Surrender, and Germany regained full sovereignty. This permitted German reunification on 3 October 1990, with the accession of the five re-established states in the former GDR (New states or "neue Lander").

Berlin Republic and EU Integration (1990)

Based on the Bonn-Berlin Act, adopted by the parliament on 10 March 1994, Berlin once again became the capital of the reunified Germany, while Bonn obtained the unique status of a *Bundesstadt* (federal city) retaining some federal ministries. The relocation of the government was completed in 1999.

Since reunification, Germany has taken a more active role in the European Union and NATO. Germany sent a peacekeeping force to secure stability in the Balkans and sent a force of German troops to Afghanistan as part of a NATO effort to provide security in that country after the ousting of the Taliban. These deployments were controversial, since after the war, Germany was bound by domestic law only to deploy troops for defence roles. Deployments to foreign territories were understood not to be covered by the defence provision; however, the parliamentary vote on the issue effectively legalised the participation in a peacekeeping context.

Geography

Germany is a country in Central Europe, stretching from the Alps, across the North European Plain to the North Sea and the Baltic Sea. Germany has the second largest population in Europe (after the European part of Russia) and is seventh largest in area. The territory of Germany covers 357,021 km (137,847 sq mi^2), consisting of 349,223 km^2 (134,836 sq mi) of land and 7,798 km^2 (3,011 sq mi) of water. Elevation ranges from the mountains of the Alps (highest point: the Zugspitze at 2,962 metres (9,718 ft) in the south to the shores of the North Sea (Nordsee) in the northwest and the Baltic Sea (Ostsee) in the northeast. Between lie the forested uplands of central Germany and the low-lying lands of northern Germany (lowest point: Neuendorf-Sachsenbande at 3.54 metres (11.6 ft) below sea level), traversed by some of Europe's major rivers such as the Rhine, Danube and Elbe. Because of its central location,

Germany shares borders with nine European countries, second only to Russia: Denmark in the north, Poland and the Czech Republic in the east, Austria and Switzerland in the south, France in the southwest and Belgium, Luxembourg and the Netherlands in the west. German Unity Day is celebrated on October 3 every year.

The current, post-1990 territory of the Federal Republic of Germany is 30.3% bigger than the Cold-War era West Germany and 23.8% and 33.9% smaller than the former territories of Weimar and Imperial Germany, respectively.

Terrain

The northern third of the country lies in the North European Plain, with flat terrain crossed by northward-flowing watercourses (Elbe, Ems, Weser, Oder). Wetlands and marshy conditions are found close to the Dutch border and along the Frisian coast. Sandy Mecklenburg in the northeast has many glacier-formed lakes dating to the last glacial period.

Moving south, central Germany features rough and somewhat patternless hilly and mountainous countryside, some of it formed by ancient volcanic activity. The Rhine valley cuts through the western part of this region. The central uplands continue east and north as far as the Saale and merge with the Ore Mountains on the border with the Czech Republic. Upland regions include the Eifel, Hunsruck and Palatine Forest west of the Rhine, the Taunus hills north of Frankfurt, the Vogelsberg massif, the Rhon, and the Thuringer Wald. South of Berlin, the east-central part of the country is more like the low northern areas, with sandy soil and river wetlands such as the Spreewald region.

Southern Germany's landforms are defined by various linear hill and mountain ranges like the two adjacent ranges of the Swabian and Franconian Alb (reaching approximately from the source of the Danube in the southwest of Baden-Wurttemberg, south of Stuttgart, across Swabia into Central Franconia and to the valley of the river Main) and the Bavarian Forest along the border between Bavaria and the Czech Republic. The Alps on the southern border are the highest mountains, but relatively little Alpine terrain lies within Germany (in southeastern Swabia and Upper Bavaria) compared to Switzerland and

Austria. The Black Forest, on the southwestern border with France, separates the Rhine from the headwaters of the Danube on its eastern slopes.

Climate

Germany's climate is temperate and marine, with cool, cloudy, wet winters and summers and in the south occasional warm fohn wind. The greater part of Germany lies in the cool/ temperate climatic zone in which humid westerly winds predominate.

In the northwest and the north the climate is extremely oceanic and rain falls all the year round. Winters there are relatively mild and summers comparatively cool. In the east the climate shows clear continental features; winters can be very cold for long periods, and summers can become very warm. Here too, long dry periods are often recorded. In the centre and the south there is a transitional climate which may be predominantly oceanic or continental, according to the general weather situation. Winters are mild and summers tend to be cool, though maximum temperatures can exceed 30 °C (86 °F) for several days in a row during heat waves. The warmest regions of Germany can be found in the south-west. Here summers can be hot with a high number of days exceeding 30 °C (86 °F). Sometimes minimum temperatures don't drop under 20 °C (68 °F), which is relatively rare in other regions.

Land Use

Germany covers a total of 357,021 km^2, of which 4,750 km^2 is irrigated land and 7,798 km^2 is covered by water, the largest lakes being Lake Constance (total area of 536 km^2, with 62% of the shore being German; international borders are not defined on the lake itself), Muritz (117 km^2) and Chiemsee (80 km^2). The majority of Germany is covered by either arable land (33%) or forestry and woodland (31%). Only 15% is covered by permanent pastures.

Germany has a total of 2,389 km of coastline, and borders totalling 3,621 km (clockwise from north: Denmark 68 km, Poland 456 km, Czech Republic 646 km, Austria 784 km, Switzerland 334 km, France 451 km, Luxembourg 138 km, Belgium 167 km, Netherlands 577 km). The German-Austrian border crosses itself near Jungholz. The border with Belgium includes 5 German exclaves because the Vennbahn railway is on Belgian territory crossing in and out of Germany.

Caves

Throughout the Karst rocks many caves were formed especially in the valley of the Honne. The biggest culture cave of Europe is located in Balve.

Rivers

The three main rivers in Germany are:

- the Rhine (*Rhein* in German) with a German part of 865 km (main tributaries including the Neckar, the Main and the Moselle (*Mosel*));
- the Elbe with a German part of 727 km (also drains into the North Sea); and,
- the Danube (*Donau*) with a German part of 687 km.

Further important rivers include the Isar in the southeast, the Main in central Germany, the Neckar in the southwest, and the Weser in the North. See also the list of rivers in Germany.

Maritime Claims

- Continental shelf: 200 m depth or to the depth of exploitation
- Exclusive economic zone: 200 nmi (370.4 km; 230.2 mi). The exact EEZ is due to conventions with neighbouring states.
- Territorial sea: 12 nmi (22.2 km; 13.8 mi).

Natural Resources

- Iron ore, coal, potash, timber, lignite, uranium, copper, natural gas, salt, nickel, arable land, water.

Environment

Current Issues

Emissions from coal-burning utilities and industries contribute to air pollution; acid rain, resulting from sulphur dioxide emissions, is damaging forests; pollution in the Baltic Sea from raw sewage and industrial effluents from rivers in eastern Germany; hazardous waste disposal; government (under Chancellor Schroder, SPD) announced intent to end the use of nuclear power for producing electricity; government working to meet EU commitment to identify nature preservation areas in line with the EU's Flora, Fauna, and Habitat directive. Germany's last glacier is disappearing.

International Agreements

Party to: Air Pollution, Air Pollution-Nitrogen oxides, Air Pollution-Sulphur 85, Air Pollution-Sulphur 94, Air Pollution-Volatile Organic Compounds, Antarctic-Environmental Protocol, Antarctic Treaty, Biodiversity, Climate Change, Climate Change-Kyoto Protocol, Desertification, Endangered Species, Environmental Modification, Hazardous Wastes, Law of the Sea, Marine Dumping, Nuclear Test Ban, Ozone Layer Protection, Ship Pollution, Tropical Timber 83, Tropical Timber 94, Wetlands, Whaling

Signed, but not ratified: Air Pollution-Persistent Organic Pollutants

Natural Hazards

Flooding through rivers after heavy rainfall, such as during the 2002 European floods, or Storm surge, such as the North Sea flood of 1962 or the historic floods of 1634 and 1362 that changed the coastline of what is now the west coast of Schleswig-Holstein.

Latitude and Longitude

This is a list of the extreme points of Germany, the points that are farther north, south, east or west than any other location.

- northernmost point: List, Sylt, Schleswig-Holstein 55°03′N 8°24′E/55.05°N 8.4°E
- southernmost point: Haldenwanger Eck, Oberstdorf, Bavaria 47°16′ 12.39′N 10°10′ 41.95′E/47.2701083°N 10.1783194°E
- westernmost point: Isenbruch, Selfkant, North Rhine-Westphalia 51°1′N 5°53′E/51.017°N 5.883°E
- easternmost point: Deschka, Neibeaue, Saxony 51°16′N 15°2′E/51.267°N 15.033°E.

The extreme points of the German Confederation are mentioned in the first stanza of *Das Lied der Deutschen*, of which the third stanza is today the national anthem of Germany. They were accurate when the song was penned in 1841, but are no longer accurate today. The limits mentioned are the following rivers:

- Meuse River which crosses France, Belgium and the Netherlands and formed the border of the Duchy of Limburg, then part of the German Confederation and today the Dutch Province of Limburg.
- Memel, which runs through Belarus and Lithuania but previously formed part of the border of East Prussia
- Adige (*Etsch*), beginning in the province of Bolzano-Bozen (then in Austria) which was transferred from the Habsburg Empire to Italy after World War I
- Belt which is a part of the Baltic Sea between Germany and Denmark.

Elevation

The territory of Germany covers 357,021 km^2 (137,847 sq mi), consisting of 349,223 km^2 (134,836 sq mi) of land and 7,798 km^2 (3,011 sq mi) of water. It is the seventh largest country by area in Europe and the 63rd largest in the world. Elevation ranges from the mountains of the Alps (highest point: the Zugspitze at 2,962 metres /9,718 feet) in the south to the shores of the North Sea (Nordsee) in the north-west and the Baltic Sea (Ostsee) in the north-east. Between lie the forested uplands of central Germany and the low-lying lands of northern Germany (lowest point: Wilstermarsch at 3.54 metres /11.6 feet below sea level), traversed by some of Europe's major rivers such as the Rhine, Danube and Elbe.

Germany shares borders with more European countries than any other country on the continent. Its neighbours are Denmark in the north, Poland and the Czech Republic in the east, Austria and Switzerland in the south, France and Luxembourg in the south-west and Belgium and the Netherlands in the north-west.

Climate

Most of Germany has a temperate seasonal climate in which humid westerly winds predominate. The climate is moderated by the North Atlantic Drift, which is the northern extension of the Gulf Stream. This warmer water affects the areas bordering the North Sea including the area along the Rhine, which flows into the North Sea. Consequently in the north-west and the north, the climate is oceanic; rainfall occurs year round with a maximum during summer.

Winters are mild and summers tend to be cool, though temperatures can exceed 30 °C (86 °F) for prolonged periods. In the east, the climate is more continental; winters can be very cold, summers can be very warm, and long dry periods are often recorded. Central and southern Germany are transition regions which vary from moderately oceanic to continental.

Biodiversity

Phytogeographically, Germany is shared between the Atlantic European and Central European provinces of the Circumboreal Region within the Boreal Kingdom. The territory of Germany can be subdivided into two ecoregions: European-Mediterranean montane mixed forests and Northeast-Atlantic shelf marine. The majority of Germany is covered by either arable land (33%) or forestry and woodland (31%). Only 15% is covered by permanent pastures.

Plants and animals are those generally common to middle Europe. Beeches, oaks, and other deciduous trees constitute one-third of the forests; conifers are increasing as a result of reforestation. Spruce and fir trees predominate in the upper mountains, while pine and larch are found in sandy soil. There are many species of ferns, flowers, fungi, and mosses. Fish abound in the rivers and the North Sea. Wild animals include deer, wild boar, mouflon, fox, badger, hare, and small numbers of beaver. Various migratory birds cross Germany in the spring and autumn.

The national parks in Germany include the Wadden Sea National Parks, the Jasmund National Park, the Vorpommern Lagoon Area National Park, the Muritz National Park, the Lower Oder Valley National Park, the Harz National Park, the Saxon Switzerland National Park and the Bavarian Forest National Park.

Germany is known for its many zoological gardens, wildlife parks, aquaria, and bird parks. More than 400 registered zoos and animal parks operate in Germany, which is believed to be the largest number in any single country of the world. The Zoologischer Garten Berlin is the oldest zoo in Germany and presents the most comprehensive collection of species in the world.

Environment

Germany is known for its environmental consciousness. Most Germans consider anthropogenic causes to be a significant factor in global warming. The state is committed to the Kyoto protocol and several other treaties promoting biodiversity, low emission standards, recycling, and the use of renewable energy, and supports sustainable development at a global level.

The German government has initiated wide-ranging emission reduction activities and the country´s overall emissions are falling. For example, since 1964, air pollution in Germany has been regulated by strict "TA Luft" legislation. Nevertheless Germany's carbon dioxide emissions per capita are among the highest in the EU, although they are significantly lower than those of Australia, Canada, Saudi Arabia and the United States.

Emissions from coal-burning utilities and industries contribute to air pollution. Acid rain, resulting from sulphur dioxide emissions, continues to damage German forests. Pollution in the Baltic Sea from raw sewage and industrial effluents from rivers in former East Germany have been reduced. The government under Chancellor Schroder announced the intention to end the use of nuclear power for producing electricity. Germany is working to meet EU commitments to identify nature preservation areas in line with the EU's Flora, Fauna, and Habitat directive. Germany's last glaciers in the Alpine region are experiencing deglaciation. Natural hazards are river flooding in spring and stormy winds occurring in all regions.

Government

Germany is a federal, parliamentary, representative democratic republic. The German political system operates under a framework laid out in the 1949 constitutional document known as the *Grundgesetz* (Basic Law). By calling the document *Grundgesetz*, rather than *Verfassung* (constitution), the authors expressed the intention that it would be replaced by a proper constitution once Germany was reunited as one state. Amendments to the *Grundgesetz* generally require a two-thirds majority of both chambers of the parliament; the fundamental principles of the constitution, as expressed in the articles guaranteeing human dignity, the separation of powers, the federal structure, and the rule of law are valid in perpetuity. Despite the initial intention, the *Grundgesetz* remained in effect after the German reunification in 1990, with only minor amendments.

The Chancellor—currently Angela Merkel—is the head of government and exercises executive power, similar to the role of a Prime Minister in other parliamentary democracies. Federal legislative power is vested in the parliament consisting of the *Bundestag* (Federal Diet) and *Bundesrat* (Federal Council), which together form a unique type of legislative body. The *Bundestag* is elected through direct elections, by proportional representation (mixed-member). The members of the *Bundesrat* represent the governments of the sixteen federal states and are members of the state cabinets. The respective state governments have the right to appoint and remove their envoys at any time.

The President—currently vacant, acting head of state is Jens Bohrnsen—is the head of state, invested primarily with representative responsibilities and powers. He is elected by the *Bundesversammlung* (federal convention), an institution consisting of the members of the *Bundestag* and an equal number of state delegates. The second highest official in the German order of precedence is the *Bundestags president* (President of the *Bundestag*), who is elected by the *Bundestag* and responsible for overseeing the daily sessions of the body. The third-highest official and the head of government is the Chancellor, who is nominated by the *Bundesprasident* after being elected by the *Bundestag*. The Chancellor can be removed by a constructive motion of no confidence by the *Bundestag*, where constructive implies that the *Bundestag* simultaneously elects a successor.

Since 1949, the party system has been dominated by the Christian Democratic Union and the Social Democratic Party of Germany with all chancellors hitherto being member of either party. However, the smaller liberal Free Democratic Party (which has had members in the *Bundestag* since 1949) and the Alliance '90/The Greens (which has controlled seats in parliament since 1983) have also played important roles, as they are regularly the smaller partner of a coalition government.

Law

The Judiciary of Germany is independent of the executive and the legislative branches. Germany has a civil or statute law system that is based on Roman law with some references to Germanic law. The *Bundesverfassungsgericht* (Federal Constitutional Court), located in Karlsruhe, is the German Supreme Court responsible for constitutional matters, with power of judicial review. It acts as the highest legal authority and ensures that legislative and judicial practice conforms to the Basic Law for the Federal Republic of Germany (Basic Law). It acts independently of the other state bodies, but cannot act on its own behalf.

Germany's supreme court system, called *Oberste Gerichtshofe des Bundes*, is specialised. For civil and criminal cases, the highest court of appeal is the Federal Court of Justice, located in Karlsruhe and Leipzig. The courtroom style is inquisitorial. Other Federal Courts are the Federal Labour Court in Erfurt, the Federal Social Court in Kassel, the Federal Finance Court in Munich and the Federal Administrative Court in Leipzig.

Criminal law and private law are codified on the national level in the *Strafgesetzbuch* and the *Burgerliches Gesetzbuch* respectively. The German penal system is aimed towards rehabilitation of the criminal; its secondary goal is the protection of the general public. To achieve the latter, a convicted criminal can be put in preventive detention (*Sicherungsverwahrung*) in addition to the regular sentence if he is considered to be a threat to the general public. The *Völkerstrafgesetzbuch* regulates the consequences of crimes against humanity, genocide and war crimes. It gives German courts universal jurisdiction if prosecution by a court of the country where the crime was committed, or by an international court, is not possible.

State Legislation

Legislative power is divided between the federation and the state level. The Basic Law presumes that all legislative power remains at the state level unless otherwise designated by the Basic Law itself. Any federal law overrides state law if the legislative power lies at the federal level. A famous example is the purported Hessian provision for the death penalty, which goes against the ban on capital punishment under the Basic Law, rendering the Hessian provision invalid. The *Bundesrat* is the federal organ through which the states participate in national legislation. State participation in federal legislation is necessary if the law falls within the area of concurrent legislative power, requires states to administer federal regulations, or is so designated by the Basic Law. Every state has its own constitutional court. The *Amtsgerichte*, *Landgerichte* and *Oberlandesgerichte* are state courts of general jurisdiction. They are competent whether the action is based on federal or state law.

Many of the fundamental matters of administrative law remain in the jurisdiction of the states, though most states base their own laws in that area on the 1976 *Verwaltungsverfahrensgesetz* (Administrative Proceedings Act) covering important points of administrative law. The *Oberverwaltungsgerichte* are the highest level of administrative jurisdiction concerning the state administrations, unless the question of law concerns federal law or state law identical to federal law. In such cases, final appeal to the Federal Administrative Court is possible.

Foreign Relations

Germany has played a leading role in the European Union since its inception and has maintained a strong alliance with France since the end of World War II. The alliance was especially close in the late 1980s and early 1990s under the leadership of Christian Democrat Helmut Kohl and Socialist Francois Mitterrand. Germany is at the forefront of European states seeking to advance the creation of a more unified European political, defence and security apparatus. For a number of decades after WWII, the Federal Republic of Germany kept a notably low profile in international relations, because of both its recent history and its occupation by foreign powers.

During the Cold War, Germany's partition by the Iron Curtain made it a symbol of East-West tensions and a political battleground in Europe. However, Willy Brandt's Ostpolitik was a key factor in the *detente* of the 1970s. In 1999, Chancellor

Gerhard Schroder's government defined a new basis for German foreign policy by taking a full part in the decisions surrounding the NATO war against Yugoslavia and by sending German troops into combat for the first time since World War II.

The governments of Germany and the United States are close political allies. The 1948 Marshall Plan and strong cultural ties have crafted a strong bond between the two countries, although Schroder's very vocal opposition to the Iraq War suggested the end of Atlanticism and a relative cooling of German-American relations. The two countries are also economically interdependent: 8.8% of German exports are U.S.-bound and 6.6% of German imports originate from the U.S. The other way around, 8.8% of U.S. exports ship to Germany and 9.8% of U.S. imports come from Germany. Other signs of the close ties include the continuing position of German-Americans as the largest ethnic group in the U.S. and the status of Ramstein Air Base (near Kaiserslautern) as the largest U.S. military community outside the U.S.

Development Aid

The development policy of the Federal Republic of Germany is an independent area of German foreign policy. It is formulated by the Federal Ministry for Economic Cooperation and Development (BMZ) and carried out by the implementing organisations. The German government sees development policy as a joint responsibility of the international community.

Germany's official development aid and humanitarian aid for 2007 amounted to 8.96 billion euros (12.26 billion dollars), an increase of 5.9 per cent from 2006. It has become the world's second biggest aid donor after the United States. Germany spent 0.37 per cent of its gross domestic product (GDP) on development, which is below the government's target of increasing aid to 0.51 per cent of GDP by 2010. The international target of 0.7% of GNP would have not been reached either.

Military

Germany's military, the *Bundeswehr*, is a military force with *Heer* (Army), *Marine* (Navy), *Luftwaffe* (Air Force), *Zentraler Sanitatsdienst* (Central Medical Services) and *Streitkraftebasis* (Joint Support Service) branches. Military service is compulsory for men at the age of 18, and conscripts serve six-month tours of duty. Conscientious objectors may instead opt for an equal length of *Zivildienst* (roughly translated as civilian service), or a six year commitment to (voluntary) emergency services like a fire department, the Red Cross or the THW. In 2003, military spending constituted 1.5% of the country's GDP. In peacetime, the Bundeswehr is commanded by the Minister of Defence, currently Karl-Theodor zu Guttenberg. If Germany went to war, which according to the constitution is allowed only for defensive purposes, the Chancellor would become commander in chief of the *Bundeswehr*.

The Bundeswehr employs 200,500 professional soldiers, 55,000 18–25 year-old conscripts who serve for at least six months under current rules, and 2,500 active reservists at any given time. Roughly 300,000 reservists are available to the Armed Forces and participate in defense exercises as well as deployments abroad. Since 2001 women can serve in all functions of service without restriction, but they are not subject to conscription. There are presently around 14,500 women on active duty and a number of female reservists who take part in all duties including peacekeeping missions and other operations. Two female medical officers have been promoted to a General rank so far.

As of November 2009, the German military had about 8,300 troops stationed in foreign countries as part of various international peacekeeping forces, including 2,470 Bundeswehr soldiers in Kosovo, 4,520 German troops in the NATO-led ISAF force in Afghanistan and Uzbekistan, and 450 troops with UNIFIL in Lebanon.

In 2009, Defence Minister Karl-Theodor zu Guttenberg stated that conditions in Afghanistan were "like a war", while it previously had been referred to as "stabilisation and civilian reconstruction", avoiding the word "war".

Demographics

With 82 million inhabitants in January 2010, Germany is the most populous country in the European Union. However, its fertility rate of 1.38 children per mother is one of the lowest in the world, and the federal statistics office estimates the population will shrink to between 65 and 70 million by 2060

(65 million assuming a net migration of +100,000 per year; 70 million assuming a net migration of +200,000 per year). With death rates continuously exceeding low-level birth rates, Germany is one of a few countries for which the demographic transition model would require a fifth stage in order to capture its demographic development. Germany has a number of large cities; the most populous are: Berlin, Hamburg, Munich, Cologne, Frankfurt, and Stuttgart. By far the largest conurbation is the Rhine-Ruhr region (12 million), including Düsseldorf (the capital of North Rhine-Westphalia), Cologne, Essen, Dortmund, Duisburg, and Bochum.

Germans make up 91% of the population of Germany. As of December 2004, about seven million foreign citizens were registered in Germany, and 9% of the country's residents were of foreign or partially foreign descent. The young are more likely to be of foreign descent than the old. 30% of Germans aged 15 years and younger have at least one parent born abroad. In the big cities 60% of children aged 5 years and younger have at least one parent born abroad. As of 2008, the largest national group of people with a migrant background was from Turkey (2.5 million), followed by Italy (776,000) and Poland (687,000).

The United Nations Population Fund lists Germany as host to the third-highest number of international migrants worldwide, about 5% or 10 million of all 191 million migrants As a consequence of restrictions to Germany's formerly rather unrestricted laws on asylum and immigration, the number of immigrants seeking asylum or claiming German ethnicity (mostly from the former Soviet Union) has been declining steadily since 2000. Large numbers of people with full or significant German ancestry are found in the United States (50 million), Brazil (5 million) and Canada (3 million). About 3 million "Aussiedler"—ethnic Germans, mainly from Eastern Europe and the former Soviet Union—have resettled in Germany since 1987.

Religion

According to organizational reportings, Christianity is the largest religion in Germany, with 52.116 million adherents (63%) in 2007. 26.5 million are Protestants (32.3%) and 25.5 million are Catholics (31.0%) in 2007. The second largest religion is Islam with 4.3 million adherents (5%) followed by Buddhism and Judaism, both with around 200,000 adherents (c. 0.25%). Hinduism has some 90,000 adherents (0.1%) and Sikhism 75,000 (0.09%). All other religious communities in Germany have fewer than 50,000 (or less than 0.05%) adherents. About 24.4 million Germans (29.6%) have no registered religious denomination. However, polls and surveys put the number of atheists and agnostics at considerably higher percentages than the religious bodies do. "God is Dead" wrote German philosopher Friedrich Nietzsche in the late 19th century. Today, for the first time in history, most Germans believe him. According to a poll by Der Spiegel magazine, only 45% believe in God, and just a quarter in Jesus Christ.

Protestantism is concentrated in the north and east and Roman Catholicism is concentrated in the south and west. The current Pope, Benedict XVI, was born in Bavaria. Non-religious people, including atheists and agnostics might make as many as 55%, and are especially numerous in the former East Germany and major metropolitan areas.

Of the 4.3 million Muslims, most are Sunnis and Alevites from Turkey, but there are a small number of Shi'ites. 1.7% of the country's overall population declare themselves Orthodox Christians, Serbs and Greeks being the most numerous. Germany has Europe's third-largest Jewish population (after France and the United Kingdom). In 2004, twice as many Jews from former Soviet republics settled in Germany as in Israel, bringing the total Jewish population to more than 200,000, compared to 30,000 prior to German reunification. Large cities with significant Jewish populations include Berlin, Frankfurt and Munich. Around 250,000 active Buddhists live in Germany; 50% of them are Asian immigrants. According to the Eurobarometer Poll 2005, 47% of German citizens agreed with the statement "I believe there is a God", whereas 25% agreed with "I believe there is some sort of spirit or life force" and 25% said "I do not believe there is any sort of spirit, god, or life force".

Languages

German is the official and predominant spoken language in Germany. It is one of 23 official languages in the European Union, and one of the three working languages of the European

Commission, along with English and French. Recognised native minority languages in Germany are Danish, Sorbian, Romany, and Frisian. They are officially protected by the ECRML. The most used immigrant languages are Turkish, Kurdish, Polish, the Balkan languages, and Russian.

Standard German is a West Germanic language and is closely related to and classified alongside English, Dutch, and the Frisian languages. To a lesser extent, it is also related to the East (extinct) and North Germanic languages. Most German vocabulary is derived from the Germanic branch of the Indo-European language family. Significant minorities of words are derived from Latin and Greek, with a smaller amount from French and most recently English (known as Denglisch). German is written using the Latin alphabet. In addition to the 26 standard letters, German has three vowels with Umlauts, namely a, o, and u, as well as the Eszett or *scharfes S* (*sharp s*) which is written "ß".

German dialects are distinguished from varieties of standard German. German dialects are traditional local varieties and are traced back to the different German tribes. Many of them are not easily understandable to a speaker of standard German, since they often differ in lexicon, phonology, and syntax.

Around the world, German has approximately 100 million native speakers and also about 80 million non-native speakers. German is the main language of about 90 million people (18%) in the EU. 67% of German citizens claim to be able to communicate in at least one foreign language, 27% in at least two languages other than their own. Social Issues in Germany

The German social market economy (*soziale Marktwirtschaft*) was the economic policy ever since the Federal Republic of West Germany was founded in 1948. This policy brought about the economic miracle that rebuilt Germany from scratch after World War II to one of the largest economies in the World. However, Germany still continues to struggle with a number of social issues.

Economic Issues

Germans continue to be concerned about a relatively high level of unemployment. Unemployment rates vary by region, gender, educational attainment and ethnic group.

A growing number of Germans are poor and depend on welfare. In 2007 one in 6 children depended on welfare. That's up from only one in 75 in 1965. Poverty rates seem to vary in different states as in Bavaria only 3.9% suffer from poverty while in Berlin 15.2% of the inhabitants are poor. Families that are headed by a single parent and working class families with multiple children are most likely to be poor.

Hunger

There is a discussion going on about hunger in Germany. Reverend Bernd Siggelkow, founder of the Berlin-based soup kitchen "Die Arche", claimed that a number of German children go hungry each day. He blamed the lack of jobs, low welfare payments, and parents who were drug-addicted or mentally ill. Siggelkow has been criticized by a number of people who said there was no hunger in Germany. SPD politician and board member of the German central bank Thilo Sarrazin said it was possible to live on welfare without going hungry if one did not buy fast food, but was able to cook from scratch. He was criticized by The Left politician Heidi Knake-Werner, who said it was not right that "well-off people told poor people how to shop".

Problem Neighbourhoods

So called problem neighbourhoods ("Problemviertel") exist in Germany. Examples are Gelsenkirchen-Bismarck, Koln-Kalk or the Ihme-Zentum in Hannover. Those neighbourhoods have a high high-school-drop-out rate and children growing up in a neighbourhood like this have only 1/7 of the probability of going to college compared to a person growing up in another neighbourhood. Abuse of alcohol and drugs is common. Many people living in those neighbourhoods are what is called *a-people*. They are poor ("arm"), out-of-work ("arbeitslos") and immigrants ("Auslander"). Often those neighbourhoods were founded out of best intentions. Many districts that later became problem neighbourhoods were founded in the 1960s and 1970s when the State wanted to provide better housing for poorer persons. Big tenement buildings were build. The first tenants mostly were two-parent-families, with at least one parent working and many were happy with their neighbourhoods. But when the unemployment rate started climbing more and more people were losing their jobs. Also, families who could afford it started moving into better districts and only those who could not afford to move stayed in districts such as Hamburg-Mummelmannsberg.

Education

Illiteracy and Innumeracy

A growing number of Germans are functionally illiterate, functionally innumerate, or both. According to a study done by the University of Bremen in cooperation with the "Bundesverband Alphabetisierung e.V.", ten percent of youngsters living in Germany are illiterate and one quarter was able to understand only basic level texts. 21.6 percent of all youngsters were only able to do mathematics at or below fourth-grade-level. The percentage of illiterate and innumerate youngsters varies among ethnic groups and parents' socioeconomic status.

School Violence

A debate about school violence has been sparked in 2006 when teachers of the Rutli School wrote a letter to the senate office demanding the school to be closed down. The letter stated:

We must realize that the mood in some classes currently is marked by aggressiveness, disrespect, and ignorance towards adults ... The tendency toward violence against property is growing ... In most of the families of our students, they are the only ones getting up in the morning. For them, school is a stage and battleground for attention. The worst culprits become role models.

Religion in Education

Some German states have banned Muslim teachers from wearing headscarves in class. Students are allowed to wear headscarves. All German States have banned crosses from the classroom. However the cross is allowed in class if none of the students objects to it, but must be removed if any student objects. Generally the use of all religious symbols by teachers is prohibited in state schools (however this is not true for faith schools). This is legitimate by combining the German states' privilege of educational laws with the principle of separation of church and state, both provided for in the German federal constitution. According to this legal view, teachers in their vocational function within a state administered educational system are obliged to maintain and publicly exhibit religious neutrality when on duty. As this status of employment does not hold for pupils, whose constitutional right to religious freedom thus remains unencumbered by these provisions, this ban cannot legally be extended to them as it is in France. Students visiting a Gymnasium are required to attend religious education or ethics classes. According to the German constituition church and state are separated in Germany. Thus while classes in religion or ethics are compulsory in a Gymnasium, the student may choose which specific religion, if any, is studied.

Reunification

Issues created by reunification in 1990 have begun to narrow. While the standard of living is still higher in the western half of the country, and unemployment is considerably higher in the East, easterners now share a reasonably high standard of living. Still, even as economic issues are abating between the two formerly separate parts of the country, societal and cultural divisions persist. Stereotypes and labels such as "Jammer-Ossi" (whining easterner), "Besser-Wessi" (arrogant westerner who always thinks he or she is right), and western resentment towards the costs of unification point to continued prejudices.

Political Extremism

Since World War II, Germany has experienced intermittent turmoil from various groups. In the 1970s radical leftist terrorist organisations like the Red Army Faction engaged in a string of assassinations and kidnappings against political and business figures. Germany has also continued to struggle with far-right violence or neo-nazis which are presently on a rise, in line with the younger generation of Germans growing older. There is some debate as to whether indeed the hate crime is rising, or whether simply more arrests have been made due to increased law-enforcement efforts. The number of officially recognized violent hate crimes has risen from 759 (2003) to 776 (2005). Some have suggested that the increase in hate crime is related to the proliferation of right-wing parties, such as the NPD (National Democratic Party) in local elections.

Gender Roles and Demographics

The Role of Women

For centuries, a woman's role in German society was summed up by the three words: Kinder, Kuche, Kirche (children, kitchen, church). Throughout the twentieth century, however, women have gradually won victories in their quest for equal rights. Despite significant gains, discrimination remains in united Germany. Women are noticeably absent in the top tiers of German businesses. They only hold 9.2% of jobs in Germany's upper and middle management positions. Until 2001 women were barred from serving in combat units in the Bundeswehr, being restricted to the medical service and the administration. The first woman to become chancellor is Angela Merkel, who was elected in 2005.

A Low Birth Rate and an Aging Population

Germany has one of the lowest birth rates in the world. In 2007 its national fertility rate was 1.45 children per woman, up slightly from the 2002 rate (1.31), but still well below the replacement rate of 2.1 children per woman. (By contrast, the United States had a fertility rate of 2.09 in 2006). At the same time, Germans are living longer, with a life expectancy of 75.6 years for men and 81.3 for women. This demographic shift is already straining the country's social welfare structures and will produce further economic and social problems in the future. The Mikrozensus done in 2008 revealed that the number of children a German women aged 40 to 75 had, was closely linked to her educational achievement. In western Germany the most educated women were the most likely to be childless. 26% of those groups stated they were childless, while only 16% of those having an intermediate education, and 11% of those having compulsory education stated the same. In Eastern Germany however, only 9% of the most educated women of that age group and only 7% of those who had a intermediary education were childless, while 12% of those having only compulsory education were childless. The reason for that east-western difference is the fact that the GDR had an "educated mother scheme" and actively tried to encourage first births among the more educated.

The more educated a western German mother aged 40 to 75 is, the less likely she is to have a big family.

Percent of western German mothers having 1, 2 and 3 and more children by educational attainment:

Number of children	*Compulsory education*	*Intermediary education*	*Highest education*
one child	22	30	31
two children	39	48	48
three or more children	39	22	21

The same is true for mothers living in Eastern Germany.

Out-of-wedlock-births

In 1950, 8% of all children born in Germany were born out of wedlock and in 1990, 15% of all births were to unmarried women. There has been a sharp increase in the out-of-wedlock-rate since that time and in 2006 one in three children was born out-of-wedlock. The number of children born out-of-wedlock varies by region, ethnic group (muslims for example are very unlikely to have children out of wedlock) and social class. Children born out-of-wedlock face a number of problems, including poverty. In 2008, 43% of families headed by a single women had to rely on welfare as the main source of household income.

Teenage Pregnancy, Abortion and Teenage Parenthood

Students attending German school receive sex education classes, which typically discuss condoms and other birth control methods as well as abstinence. While teenage pregnancy rates in Germany are lower than they are in countries such as the UK or the USA, they have been of increasing interest for the media, which have published articles on "children having children". A study revealed that teenage pregnancy was more common among young girls who had been disadvantaged in childhood, did poorly in school, and had poor expectations on the job market. It was also revealed that pregnant teenagers were more likely to come from a "broken home" than teenagers who did not fall pregnant. There also seems to be a "family tradition" of teenage parenthood as 34% of the mothers of pregnant teenagers were 20 or younger when they had their first child. While only 4% of pregnant teens planned their pregnancy, 34% did not use contraception. 9.8% of all pregnant teens and 13.4% of all pregnant 17-year-olds already had a child before they became pregnant again.

A pregnant German teen has two option. She can carry the pregnancy to term or choose an abortion. 60% of pregnant teens choose to have an abortion. The higher a pregnant teen's educational level is, the higher her partner's educational level is, and the higher her parents' socioeconomic class is, the greater is her likelihood to opt for an abortion. Pregnant Catholic or Protestant teens are just as likely as atheist teens to have an abortion when they experienced unwanted pregnancy. Muslim teens are more likely to carry the pregnancy to term. Several studies revealed that a high percentage of teenaged parents were special-needs kids. Many of those special-needs teenagers find themselves unable to care for their offspring's emotional needs. A teenaged parent may ask a Familienhelfer (family helper) to help her out. Family helpers are social workers who will help the young parents run the household, go shopping for them, or help them with the kids. A young parent may also choose to move to a *Mutter-Kind-Heim* ("Mother-Child-Home"). A Mutter-Kind-Heim is a place were troubled parents live and professionals care for the young parent and her offspring.

Integration of Immigrants

Immigration continues to be a concern of both economic and social importance. Germany has always been and still is a society with a considerable rate of immigration. Together with the enactment of a new set of immigration laws, integration of migrants has become a main focus of official federal policy. By virtue of language courses and courses on culture, politics, and society, which are largely state-financed, integration of new migrants is regulated country-wide; in some specified cases, participation in such courses is compulsory. Furthermore, the new law provides for fewer formalities and more options for highly skilled third-country nationals to enter the country for working there; citizens of European Union member states generally enjoy the right to abode and work in Germany, thus, their stay is not regulated. At the same time, with a view to security threats by international terrorism, expulsion of foreign hate-mongers and suspected terrorists in Germany has been made easier. As a result of enhanced security measures, immigrants (especially those from Muslim and African countries) may face police inquiries (such as requests for identification), which some voices regard as undue.

New immigrants face prejudices and problems integrating with the native population and often segment into separate communities. Higher rates of delinquency and more general integration problems persist amongst some migrant groups. Notwithstanding police operations focusing on this matter, migrants may still be subject to racist assaults mainly in rural areas or small towns in former East Germany. This being said, such problems are not unique to Germany and the incidence of the more severe of these problems are relatively rare in perspective. Some German states (which are responsible for education affairs) have banned Muslim teachers from wearing headscarves in class, generally by prohibiting the use of all religious symbols by teachers; this issue is being intensively discussed in the general public. Unlike France, German states have not extended this ban to pupils. German political backlash due to integration with Europe is resulting in new laws seeking to equalize the pay of eastern European workers (such as those from Poland and the Czech Republic) in an effort to curb the advantages to their hire. Alleged and real competition of "cheap" labour force mainly from Eastern Europe is an issue which right and far-right political groups-in some areas with some success-try to use to promote nationalist approaches in immigration policy. However, during the last decades, no far-right movement has been able to gain enough support to win seats in federal elections or to play any important role in politics on the federal level.

A study released by the OECD revealed that immigrants in Germany perform much worse at school than their counterparts elsewhere. There were great differences in scholastic performance between different groups of immigrants.

Gay Issues

Germany has some of the most liberal laws in Europe regulating the status of gays and lesbians, though less so than other countries such as the Netherlands. Gay partnerships which are just short of marriage have been permitted since 2001. Gays and lesbians can legally adopt their partner's biological children (so called stepchild adoption). Legally partnered gays are exempted from Germany's compulsory conscription (as are married heterosexuals). Legally partnered gays are not required to testify against each other in court. That being said, gay couples do not enjoy the tax benefits that heterosexual couples do. Politically, this has caused a "clash of cultures", with the states of Bavaria, Saxony and Thuringia suing against the introduction of "life partnerships" in 2001. They argued the law was violating a clause of the Basic Law stating "marriage and family" were under the state's "special protection". In 2002, the Federal Court of Justice of Germany ruled that this protection did not mean the legislature could not grant other arrangements similar, or even the same, rights. Nevertheless, Bavaria has declared its intent to sue against the recently introduced stepchild adoption. The Christian Social Union and Christian Democratic Union of Germany political parties have been actively working to restrict extensions in legal rights granted to homosexuals. At the same time the opposition leader of the Free Democratic Party, Guido Westerwelle, has called for an increase in rights (in particular economic rights) and has stated that the government has not gone far enough (in spite of having voted against the introduction of life partnerships in parliament in 2001). In general, Germany is one of the more legally and socially tolerant countries towards homosexuals. Despite this general tolerance, debate on the status of homosexuals continues.

The Debate about the "New Underclass"

The German underclass debate was sparked by an 2004 article in the German magazine Stern and by SPD chairman Kurt Beck. In his 2004 article Walter Wullenweber addressed social issues such as child poverty, unemployment and out of wedlock births. However, unlike many scholars, he does not perceive lack of economic opportunity, but rather lack of education and the failure of underclass persons to embrace middle class values to be the roots of this social problem. In 2006 Beck stated that "There are far too many people in Germany who see no more hope of improving their situation" and "Many people call it an underclass problem."

A number of journalists have picked up that word and published articles about what they called underclass-women and described them as having "no job, no education, no chance, no husband, no love". They were perceived as unfit mothers who could not provide for their large families. Underclass males were perceived as violent and absent fathers. This portrayal of poor persons has been criticized by a number of scientists such as Fabian Kessel, who said that the media were wrong and there was no underclass. The underclass is a construct of the mass media according to Kessl

In a 2009 interview SPD politician and board member of the German central bank Thilo Sarrazin talked about ethnic minorities in Germany. According to Sarrazin most people from certain ethnic groups were of low IQ, were neither willing nor able to contribute to society, did not embrace German values, hated German society, and kept having children in order to increase their welfare benefits. Sarrazin has been criticized by politicians Vural Oger (who accused him of blaming the victim) and Gerhard Schick (who has called his opinion "nasty"). Safter Cynar, spokesman for the Turkish Association in Berlin, said Sarrazin's remarks were insulting and untrue.

Economy

Germany is the largest national economy in Europe, the fourth-largest by nominal GDP in the world, and fifth by GDP (PPP) in 2008. Since the age of industrialisation, the country has been a driver, innovator, and beneficiary of an ever more globalised economy. Germany is the world's second largest exporter with $1.170 trillion exported in 2009 (Eurozone countries are included). Exports account for more than one-third of national output.

Germany is relatively poor in raw materials. Only lignite and potash salt are available in economically significant quantities. Power plants burning lignite are one of the main sources of electricity in Germany. Oil, natural gas and other resources are, for the most part, imported from other countries. Germany imports about two thirds of its energy.

The service sector contributes around 70% of the total GDP, industry 29.1%, and agriculture 0.9%. Most of the country's products are in engineering, especially in automobiles, machinery, metals, and chemical goods. Germany is the leading producer of wind turbines and solar power technology in the world. The largest annual international trade fairs and congresses are held in several German cities such as Hanover, Frankfurt, and Berlin.

Of the world's 500 largest stock market listed companies measured by revenue, the Fortune Global 500, 37 companies are headquartered in Germany. In 2007 the ten biggest were Daimler, Volkswagen, Allianz (the most profitable company), Siemens (one of Europe's largest employers), Deutsche Bank (2nd most profitable company), E.ON, Deutsche Post, Deutsche Telekom, Metro, and BASF. Among the largest employers are also Deutsche Post, Robert Bosch GmbH, and Edeka. Well known global brands are Mercedes Benz, SAP, BMW, Adidas, Audi, Porsche, Volkswagen, and Nivea.

As of September 2008, as measured by ILO standards the German unemployment rate was 6.2 percent (compared with 7.4 percent as measured by German standards).

History

Nazi Era

The economy of Germany during the Hitler era (1933 – 1945) developed a hothouse prosperity, supported with high government subsidies to those sectors that Hitler favoured because they gave Nazi Germany military power and economic autarky, that is, economic independence from the global economy.

Adolf Hitler, believing that "the economy is something of secondary importance", left the details of the economic National Socialist Programme out of *Mein Kampf*. The Nazis rose to power while unemployment was very high, but achieved full employment later thanks to massive rearmament. Their pre-war economic policies, resembling Keynesianism, were in the beginning the brainchildren of their non-Nazi Minister of Economics, Hjalmar Schacht, who was later made to focus more on war production (cf: Military Keynesianism), and was eventually replaced by a Nazi, Hermann Goring.

The trading policies of the Third Reich aimed at discouraging trade with countries outside the German sphere of influence, while making southern Europe largely dependent on Germany. Eventually, the Nazi party developed strong relationships with big business and abolished trade unions while real wages dropped by a fourth, and employees could not easily change employer. Taxes, though, were still low well into the war. Already before the war, people undesirable to the regime were used as slave labour, and in 1944 they reached one quarter of the workers.

Wirtschaftswunder of the West

Beginning with the replacement of the Reichsmark with the Deutsche Mark as legal tender, a lasting period of low inflation and rapid industrial growth was overseen by the government led by German Chancellor Konrad Adenauer and his minister of economics, Ludwig Erhard, raising West Germany from total wartime devastation to one of the most developed nations in modern Europe.

Contrary to popular belief, the Marshall Plan, which was extended to also include Western Germany after it was realized that the suppression of the Western German economy was holding back the recovery of the rest of Europe, was not the main force behind the *Wirtschaftswunder*.

Nonetheless, the amount of monetary aid (which was in the form of loans) received by Germany through the Marshall Plan (about $1.65 billion in total) was far overshadowed by the amount the Germans had to pay back as war reparations and by the charges the Allies made on the Germans for the ongoing cost of occupation (about $2.4 billion per year). In 1953 it was decided that Germany was to repay $1.1 billion of the aid it had received. The last repayment was made in June 1971.

Apart from these factors, hard work and long hours at full capacity among the population in the 1950s, 1960s and early 1970s and extra labour supplied by thousands of Gastarbeiter ("guest workers") provided a vital base for the economic upturn.

East Germany

By the early 1950s the Soviet Union had seized reparations in form of agricultural and industrial products and demanded further heavy reparation payments. Lower Silesia, which contained coal mines, and Stettin, a prominent natural port, were lost to Poland. Exports from West Germany exceeded $323 billion in 1988. In the same year, East Germany exported $30.7 billion of goods; 65% to other communist states. East Germany officially had zero unemployment.

In 1976 average annual GDP growth was roughly 5.9%.

Post-reunification

The German economy practically stagnated in the beginning of the 2000s. The worst growth figures were achieved in 2002 (+1.4%), in 2003 (+1.0%) and in 2005 (+1.4%). Unemployment was also chronically high. Due to these problems, together with Germany's aging population, the welfare system came under a lot of strain. This led the government to push through a wide-ranging programme of belt-tightening reforms, Agenda 2010, including the labour market reforms known as Hartz I-IV. In the latter part of the first decade of 2000 the world economy experienced high growth, from which Germany as a leading exporter also profited. The Hartz reforms contributed to the high German growth and declining unemployment, which hit a new post-reunification low of 7.8% in 2008. This period of growth came to a sudden end with the onset of the late 2000s recession.

2008–2009 Recession

The nominal GDP of Germany contracted in the second and third quarters of 2008, putting the country in a technical recession following a global and European recession cycle. German industrial output dropped to 3.6% in September vis-a-vis August. In January 2009 the German government under Angela Merkel approved a €50 billion ($70 billion) economic stimulus plan to protect several sectors from a downturn and a subsequent rise in unemployment rates.

Germany exited the recession in the second and third quarters of 2009, mostly due to rebounding manufacturing orders and exports-primarily from outside the Euro Zone-and relatively steady consumer demand.

Economic Regions

In several European countries, such as United Kingdom and France, the capital city completely dominates the national economy. Germany, on the other hand, does not have a single economic center: it is a polycentric country. Only 3 of Germany's 100 largest companies are headquartered in the capital Berlin. For example, the stock exchange is located in Frankfurt am Main, the largest Media company (Bertelsmann AG) is headquartered in Gutersloh; the most important car manufacturers are in Wolfsburg, Stuttgart and Munchen.

Old Bundesländer

One of Germany's strongest (and at the same time oldest) economic regions is the Ruhr area in the west, between Bonn and Dortmund. 27 of the country's 100 largest companies are located there. The region also has one of the highest GDP per capita figures in Germany. In recent years, however, the area, whose economy is based on natural resources and heavy industry, has seen a substantial rise in unemployment. The economy of Bayern, the state with the lowest number of unemployed people, on the other hand, is based on high-value products. Important sectors are electronics, aerospace and biomedicine, among others. The reason for the low unemployment is that Bayern started its economic rise later, after the Second World War, and does not have as many traditional industries, which are presently encountering problems due to competition from countries such as China and India.

New Bundesländer

With unification on October 3, 1990, Germany began the major task of bringing the standard of living of Germans in the former German Democratic Republic up to that of western Germany. Since reunification, hundreds of thousands of former East Germans have migrated into western Germany to find work. Drastic changes in the socioeconomic landscape brought about by reunification have resulted in troubling social problems. Economic uncertainty in eastern Germany is often cited as one factor contributing to extremist violence, primarily from the political right. Confusion about the causes of the current hardships and a need to place blame have found expression in harassment and violence by some Germans directed toward foreigners, particularly non-Europeans.

Even after the German reunification in 1990, the standard of living and annual income remains significantly higher in the former West German states. The modernisation and integration of the eastern German economy continues to be a long-term process scheduled to last until the year 2019, with annual transfers from west to east amounting to roughly $80 billion. The overall unemployment rate has consistently fallen since 2005 and reached a 15-year low in June 2008 with 7.5%. The percentage ranges from 6.2% in former West Germany to 12.7% in former East Germany.

Natural Resources

The German soil is relatively poor in raw materials. Only lignite (brown coal) and potash salt (Kalisalz) are available in significant quantities. Oil, natural gas and other resources are, for the most part, imported from other countries.

The potash salt deposits are a result of the drying up of the Zechstein sea, which 250 million years ago covered large parts of North and Central Europe. Potash salt is mined in the center of the country (Niedersachsen, Sachsen-Anhalt and Thuringen). The most important producer is K+S AG (formerly Kali und Salz AG). Germany's bituminous coal deposits were created more than 300 million years ago from swamps which extended from the present-day South England, over the Ruhr area to Poland. Lignite deposits developed in a similar way, but during a later period, about 65 million years ago. Due to the fact that the wood is not yet completely transformed into coal, brown coal contains less energy than bituminous coal.

Lignite is extracted in the extreme western and eastern pars of the country, mainly in Nordrhein-Westfalen, Sachsen and Brandenburg. Considerable amounts are burned in coal plants near to the mining areas, to produce electricity. Transporting lignite over far distances is not economically feasible, therefore the plants are located practically next to the extraction sites. Bituminous coal is mined in Nordrhein-Westfalen and Saarland. Most power plants burning bituminous coal operate on imported material, therefore the plants are located not only near to the mining sites, but throughout the country.

Sectors

Primary: In 2008 agriculture, forestry, and mining accounted for only 0.9% of Germany's gross domestic product

(GDP) and employed only 2.4% of the population, down from 4% in 1991. Much of the reduction in employment occurred in the eastern states, where the number of agricultural workers declined by as much as 75% following reunification. However, agriculture is extremely productive, and Germany is able to cover 90% of its nutritional needs with domestic production. In fact, Germany is the third largest agricultural producer in the European Union after France and Italy. Germany's principal agricultural products are potatoes, wheat, barley, sugar beets, fruit, and cabbages. Despite Germany's high level of industrialization, almost one-third of its territory is covered by forest. The forestry industry provides for about two-thirds of domestic consumption of wood and wood products, so Germany is a net importer of these items.

Industry

Industry and construction accounted for 29% of gross domestic product in 2008, and employed 29.7% of the workforce. Germany excels in the production of automobiles, machine tools, and chemicals. With the manufacture of 5.5 million vehicles in 2003, Germany was the world's third largest producer of automobiles after the United States and Japan, although the People's Republic of China was threatening to displace Germany in the world rankings as early as 2005. In 2004 Germany enjoyed the largest world market share in machine tools (19.3%). Of vital importance is the role of small-to medium-sized manufacturing firms, which specialize in niche products and often are owned by management.

Tertiary Sector

In 2008 services constituted 69% of gross domestic product (GDP), and the sector employed 67.5% of the workforce. The subcomponents of services are financial, renting, and business activities (30.5%); trade, hotels and restaurants, and transport (18%); and other service activities (21.7%).

Oil and Gas Transport

There are 3 major entry points for oil pipelines: in the northeast (the Druzhba pipeline, coming from Gdansk), west (coming from Rotterdam) and southeast (coming from Nelahozeves). The oil pipelines of Germany do not constitute a proper network, and sometimes only connect two different locations. Major oil refineries are located in or near the following cities: Schwedt, Spergau, Vohburg, Burghausen, Karlsruhe, Koln, Gelsenkirchen, Lingen, Wilhelmshafen, Hamburg and Heide.

Germany's network of natural gas pipelines, on the other hand, is dense and well-connected. Imported pipeline gas comes mostly from Russia, the Netherlands and the United Kingdom. Although gas imports from Russia have been historically reliable, even during the cold war, recent price disputes between Gazprom and the former Soviet states, such as Ukraine, have also affected Germany. As a result, high political importance is placed on the construction of the Nord Stream pipeline, running from Vyborg in Russia along the Baltic sea to Greifswald in Germany. This direct connection avoids third-party transit countries.

Problems

Germany's industrial sector is facing considerable problems due to competition from cheap labour countries such as China and India. After 1991, the country has lost about 28% its manufacturing jobs. The number of industrial employees in Germany is now lower than it was in West Germany alone before the unificiation. Labour costs in Germany are one of the highest in Europe. The disappearance of industrial jobs has created a lower class of unemployed, uneducated people who have little chance of re-entering the job market. The service sector can absorb some of those who lose their jobs, but not all.

Another problem, pointed out by Gabor Steingart, is that the country's investment ratio (investment/GDP) has sunken from 18% in 1970 to 3%, and is now only a third of that of the United States. In Western Europe (including Germany) no new company has made it to the top 100 in the last 20 years. This is in stark contrast to the United States, where newly founded companies such as eBay, Biogen, Google and Yahoo have experienced a spectacular rise.

Germany has the largest national economy in Europe, the fourth largest by nominal GDP in the world, and ranked fifth by GDP (PPP) in 2008. Since the age of industrialisation, the country has been a driver, innovator, and beneficiary of an ever more globalised economy. Germany was the world's largest exporter from 2003 to 2008. It was surpassed by China in 2009 and is currently the second largest exporter and generates a large trade surplus. The service sector contributes around 70% of the total GDP, industry 29.1%, and agriculture 0.9%. Most of the country's products are in engineering, especially in automobiles, machinery, metals, and chemical goods. Germany is the leading producer of wind turbines and solar power technology in the world. The largest annual international trade fairs and congresses are held in several German cities such as Hanover, Frankfurt, and Berlin.

Of the world's 500 largest stock market listed companies measured by revenue, the Fortune Global 500, 37 are headquartered in Germany. In 2007 the ten largest were Daimler, Volkswagen, Allianz (the most profitable company), Siemens, Deutsche Bank (2nd most profitable company), E.ON, Deutsche Post, Deutsche Telekom, Metro, and BASF. Among the largest employers are also Deutsche Post, Robert Bosch GmbH, and Edeka. Well known global brands are Mercedes Benz, SAP, BMW, Adidas, Audi, Porsche, Volkswagen, and Nivea. It is estimated that German companies were losing about €50 billion ($87 billion) and 30,000 jobs to industrial espionage every year, reportedly sponsored by China and Russia.

Germany is a strong advocate of closer European economic and political integration, and its commercial policies are increasingly determined by agreements among European Union (EU) members and EU single market legislation. Germany uses the common European currency, the euro, and its monetary policy is set by the European Central Bank in Frankfurt. Prior to 1999, the official currency was the Deutsche Mark. As of 1 January 1999, this was converted to the euro at an exchange rate of 1 euro for 1.95583 German marks, for accounting purposes. Actual euro coins and banknotes followed on 1 January 2002.

Two decades after German reunification, standards of living and per capita incomes remain significantly higher in the states of the former West Germany than in the former East. The modernisation and integration of the eastern German economy continues to be a long-term process scheduled to last until the year 2019, with annual transfers from west to east amounting to roughly $80 billion. The overall unemployment rate has consistently fallen since 2005 and reached a 15-year low in June 2008 with 7.5%. In 2009 the unemployment rate was 8% in the whole of Germany; in the former West Germany it was half the rate compared to the east.

The nominal GDP of Germany contracted in the second and third quarters of 2008, putting the country in a technical recession following a global and European recession cycle. In January 2009 the German government under Angela Merkel approved a €50 billion ($70 billion) economic stimulus plan to protect several sectors from a downturn and a subsequent rise in unemployment rates.

Infrastructure

With its central position in Europe, Germany is an important transportation hub. This is reflected in its dense and modern transportation networks. The extensive motorway (Autobahn) network that ranks worldwide third largest in its total length and features a lack of blanket speed limits on the majority of routes.

Germany has established a polycentric network of high-speed trains. The Inter City Express or *ICE* is the most advanced service category of the Deutsche Bahn and serves major German cities as well as destinations in neighbouring countries. The train maximum speed varies between 160 km/h and 300 km/h. Connections are offered at either 30-minute, hourly, or two-hourly intervals.

Germany is the world's fifth largest consumer of energy, and two-thirds of its primary energy was imported in 2002. In the same year, Germany was Europe's largest consumer of electricity, totaling 512.9 terawatt-hours. Government policy promotes energy conservation and the development of renewable energy sources, such as solar, wind, biomass, hydroelectric, and geothermal energy. As a result of energy-saving measures, energy efficiency has been improving since the beginning of the 1970s. The government has set the goal of meeting half the country's energy demands from renewable sources by 2050.

In 2000, the government and the German nuclear power industry agreed to phase out all nuclear power plants by 2021. Renewable energy still plays a more modest role in energy consumption. In 2006, energy consumption was met by the following sources: oil (36%); coal, including lignite (24%); natural gas (23%); nuclear (13%); hydro and wind power (1%); and other (4%). However, the share of renewable energy in electricity supply has been rapidly increasing, reaching 14% in 2007. The German government has set a new target to increase this share to 27% by 2020.

Science

Germany has been the home of some of the most prominent researchers in various scientific fields. The Nobel Prize has been awarded to 103 German laureates. The work of Albert Einstein and Max Planck was crucial to the foundation of modern physics, which Werner Heisenberg and Max Born developed further. They were preceded by physicists such as Hermann von Helmholtz, Joseph von Fraunhofer, and Gabriel Daniel Fahrenheit. Wilhelm Conrad Rontgen discovered X-rays, which are called *Rontgenstrahlen* (Rontgen-rays) in German and many other languages. This accomplishment made him the first winner of the Nobel Prize in Physics in 1901.

Aerospace engineer Wernher von Braun developed the first space rocket and later on was a prominent member of NASA and developed the Saturn V Moon rocket, which paved the way for the success of the US Apollo program. Heinrich Rudolf Hertz's work in the domain of electromagnetic radiation was pivotal to the development of modern telecommunication. Through his construction of the first laboratory at the University of Leipzig in 1879, Wilhelm Wundt is credited with the establishment of psychology as an independent empirical science. Alexander von Humboldt's work as a natural scientist and explorer was foundational to biogeography.

Numerous significant mathematicians were born in Germany, including Carl Friedrich Gauss, David Hilbert, Bernhard Riemann, Gottfried Leibniz, Karl Weierstrass and Hermann Weyl. Germany has been the home of many famous inventors and engineers, such as Johannes Gutenberg, who is credited with the invention of movable type printing in Europe; Hans Geiger, the creator of the Geiger counter; and Konrad Zuse, who built the first fully automatic digital computer. German inventors, engineers and industrialists such as Count

Ferdinand von Zeppelin, Otto Lilienthal, Gottlieb Daimler, Rudolf Diesel, Hugo Junkers and Karl Benz helped shape modern automotive and air transportation technology.

Important research institutions in Germany are the Max Planck Society, the Helmholtz-Gemeinschaft and the Fraunhofer Society. They are independently or externally connected to the university system and contribute to a considerable extent to the scientific output. The prestigious Gottfried Wilhelm Leibniz Prize is granted to ten scientists and academics every year. With a maximum of €2.5 million per award it is one of highest endowed research prizes in the world.

Education

Responsibility for educational oversight in Germany lies primarily with the federal states individually, whilst the federal government only has a minor role. Optional kindergarten education is provided for all children between three and six years old, after which school attendance is compulsory for at least nine years. Primary education usually lasts for four years and public schools are not stratified at this stage. In contrast, secondary education includes three traditional types of schools based on a pupil's ability as determined by teacher recommendations: the *Gymnasium* enrolls the most gifted children and prepares students for university studies, and attendance lasts eight or nine years depending on the state; the *Realschule* has a broader range of emphasis for intermediate students and lasts six years; the *Hauptschule* prepares pupils for vocational education.

Since the 1960s, a reform movement attempted to unify secondary education in a *Gesamtschule* (comprehensive school). However, instead of overcoming the stratification, Gesamtschule just became a fourth type of secondary school. Since about 2000, several West German Lander simplified their school system to two or three tiers. Motives were: The example of Eastern Germany where in the 1990s, following reunification, a two-tier school system was established; mediocre scores in the Programme for International Student Assessment (PISA), first published in 2001, prompted a nation-wide debate about the school system, and in particular about the social selectivity of early stratification; having low standards of achievement, inner-city Hauptschulen were increasingly considered dysfunctional by some people; outside the metropoles, the population is shrinking, so that according to some people it becomes increasingly unpractical to maintain a three-or four-tier school system.

A special system of apprenticeship called *Duale Ausbildung* ("dual education") allows pupils in vocational training to learn in a company as well as in a state-run vocational school.

To enter a university in Germany, high school students are generally required to take the *Abitur* examination, which is similar to A-levels in the UK and typically done at the age of 18 or 19. However, students possessing a diploma from a vocational school may also apply for matriculation in certain subjects. Germany's universities are recognised internationally, indicating the high education standards in the country. In the ARWU ranking for 2008, six of the top 100 universities in the world are in Germany, and 18 in the top 200. Nearly all German universities are public (i.e. non-private) institutions, charging tuition fees ranging from €50–500 per semester for each student.

Culture

Germany is historically called *Das Land der Dichter und Denker* (the land of poets and thinkers). German culture began long before the rise of Germany as a nation-state and spanned the entire German-speaking world. From its roots, culture in Germany has been shaped by major intellectual and popular currents in Europe, both religious and secular. As a result, it is difficult to identify a specific German tradition separated from the larger framework of European high culture. Another consequence of these circumstances is the fact that some historical figures, such as Wolfgang Amadeus Mozart, Franz Kafka and Paul Celan, though not citizens of Germany in the modern sense, must be considered in the context of the German cultural sphere in order to understand their historical situation, work and social relations.

In Germany, the Federal States are in charge of the cultural institutions. There are 240 subsidised theatres, hundreds of symphonic orchestras, thousands of museums and over 25,000 libraries spread over the 16 states. These cultural opportunities are enjoyed by many millions: there are over 91 million German museum visits every year; annually, 20 million go to theatres and operas; while 3.6 million listen to the great symphonic orchestras.

Germany claims some of the world's most renowned classical music composers, including Ludwig van Beethoven, Johann Sebastian Bach, Johannes Brahms and Richard Wagner. As of 2006, Germany is the fifth largest music market in the world and has influenced pop and rock music through artists such as Tokio Hotel, Kraftwerk, Scorpions and Rammstein.

Numerous German painters have enjoyed international prestige through their work in diverse artistic styles. Hans Holbein the Younger, Matthias Grunewald, and Albrecht Dürer were important artists of the Renaissance, Caspar David Friedrich of Romanticism, and Max Ernst of Surrealism. Architectural contributions from Germany include the Carolingian and Ottonian styles, which were important precursors of Romanesque. The region later became the site of significant works in styles such as Gothic, Renaissance and Baroque. Germany was particularly important in the early modern movement, especially through the Bauhaus movement founded by Walter Gropius. Ludwig Mies van der Rohe, also from Germany, became one of the world's most renowned architects in the second half of the 20th century. The glass facade skyscraper was his idea.

German literature can be traced back to the Middle Ages and the works of writers such as Walther von der Vogelweide and Wolfram von Eschenbach. Various German authors and poets have won great renown, including Johann Wolfgang von Goethe and Friedrich Schiller. The collections of folk tales published by the Brothers Grimm popularised German folklore on an international level. Influential authors of the 20th century include Thomas Mann, Berthold Brecht, Hermann Hesse, Heinrich Boll, and Gunter Grass.

Philosophy

Germany's influence on philosophy is historically significant and many notable German philosophers have helped shape Western philosophy since the Middle Ages. Gottfried Leibniz's contributions to rationalism; the establishment of classical German idealism by Immanuel Kant, Johann Gottlieb Fichte,

Georg Wilhelm Friedrich Hegel and Friedrich Wilhelm Joseph Schelling; Arthur Schopenhauer's composition of metaphysical pessimism; Karl Marx's and Friedrich Engels' formulation of communist theory; Friedrich Nietzsche's development of perspectivism; Gottlob Frege's contibutions to the dawn of analytic philosophy; Martin Heidegger's works on Being; and the social theories of Jurgen Habermas have been especially influential.

Media

Germany's television market is the largest in Europe, with some 34 million TV households. The many regional and national public broadcasters are organised in line with the federal political structure. Around 90% of German households have cable or satellite TV, and viewers can choose from a variety of free-to-view public and commercial channels. Pay-TV services have not become popular or successful while public TV broadcasters ZDF and ARD offer a range of digital-only channels.

Germany is home to some of the world's largest media conglomerates, including Bertelsmann and the Axel Springer AG. Some of Germany's top free-to-air commercial TV networks are owned by Pro Siebensat.

The German book market produces around 60,000 new publications every year. It represents 18% of all the books published worldwide and puts Germany in third place among the world's book producers. The Frankfurt Book Fair is considered to be the most important book fair in the world for international deals and trading and has a tradition that spans over 500 years.

The country's news is provided for English speakers e.g. by news magazine Der Spiegel and state broadcaster Deutsche Welle.

Cinema

German cinema dates back to the very early years of the medium with the work of Max Skladanowsky. It was particularly influential during the years of the Weimar Republic with German expressionists such as Robert Wiene and Friedrich Wilhelm Murnau. Austrian-based director Fritz Lang, who became a German citizen in 1926 and whose career flourished in the pre-war German film industry, is said to have been a major influence on Hollywood cinema. His silent movie Metropolis (1927) is referred to as the birth of modern Science Fiction movies. In 1930 Austrian-American Josef von Sternberg directed *The Blue Angel*, which was the first major German sound film and it brought world fame to actress Marlene Dietrich. Impressionist documentary Berlin: Symphony of a Great City, directed by Walter Ruttmann, is a prominent example of the city symphony genre. The Nazi era produced mostly propaganda films although the work of Leni Riefenstahl still introduced new aesthetics to film.

During the 1970s and 80s, New German Cinema directors such as Volker Schlondorff, Werner Herzog, Wim Wenders, and Rainer Werner Fassbinder put West German cinema back on the international stage with their often provocative films. More recently, films such as *Good Bye Lenin!* (2003), *Gegen die Wand (Head-on)* (2004), *Der Untergang (Downfall)* (2004), and *Der Baader Meinhof Komplex* (2008) have enjoyed international success. The Academy Award for Best Foreign Language Film went to the German production *Die Blechtrommel (The Tin Drum)* in 1979, to *Nowhere in Africa* in 2002, and to *Das Leben der Anderen (The Lives of Others)* in 2007. Among the most famous German actors are Marlene Dietrich, Klaus Kinski, Hanna Schygulla, Armin Mueller-Stahl, Jurgen Prochnow, and Thomas Kretschmann.

The Berlin Film Festival, held annually since 1951, is one of the world's foremost film festivals. An international jury places emphasis on representing films from all over the world and awards the winners with the Golden and Silver Bears. The annual European Film Awards ceremony is held every second year in the city of Berlin, where the European Film Academy (EFA) is located. The Babelsberg Studios in Potsdam are the oldest large-scale film studios in the world and a centre for international film production.

Sports

Sport forms an integral part of German life. Twenty-seven million Germans are members of a sports club and an additional twelve million pursue such an activity individually. Association football is the most popular sport. With more than 6.3 million official members, the German Football Association (*Deutscher*

Fußball-Bund) is the largest sports organisation of its kind worldwide. The Bundesliga attracts the second highest average attendance of any professional sports league in the world. The German national football team won the FIFA World Cup in 1954, 1974 and 1990 and the UEFA European Football Championship in 1972, 1980 and 1996. Germany has hosted the FIFA World Cup in 1974 and 2006 and the UEFA European Football Championship in 1988. Among the most successful and renowned footballers are Franz Beckenbauer, Gerd Muller, Jurgen Klinsmann, Lothar Matthaus, and Oliver Kahn. Other popular spectator sports include handball, volleyball, basketball, ice hockey, and tennis.

Germany is one of the leading motor sports countries in the world. Race winning cars, teams and drivers have come from Germany. The most successful Formula One driver in history, Michael Schumacher, has set the most significant motorsport records during his career and won more Formula One championships and races than any other driver since Formula One's debut season in 1946. He is one of the highest paid sportsmen in history and became a billionaire athlete. Constructors like BMW and Mercedes are among the leading teams in motorsport sponsorship. Porsche has won the 24 hours of Le Mans, a prestigious annual race held in France, 16 times. The Deutsche Tourenwagen Masters is a popular series in Germany. Historically, German sportsmen have been some of the most successful contenders in the Olympic Games, ranking third in an all-time Olympic Games medal count, combining East and West German medals. In the 2008 Summer Olympics, Germany finished fifth in the medal count, while in the 2006 Winter Olympics they finished first. Germany has hosted the Summer Olympic Games twice, in Berlin in 1936 and in Munich in 1972. The Winter Olympic Games took place in Germany once in 1936 when they were staged in the Bavarian twin towns of Garmisch and Partenkirchen.

Cuisine

German cuisine varies from region to region. The southern regions of Bavaria and Swabia, for instance, share a culinary culture with Switzerland and Austria. Pork, beef, and poultry are the main varieties of meat consumed in Germany, with pork being the most popular. Throughout all regions, meat is often eaten in sausage form. More than 1500 different types of sausage are produced in Germany. Organic food has gained a market share of around 3.0%, and is predicted to increase further. A popular German saying has the meaning: “Breakfast like an emperor, lunch like a king, and dine like a beggar.” Breakfast is usually a selection of breads and rolls with jam and honey or cold meats and cheese, sometimes accompanied by a boiled egg. Cereals or muesli with milk or yoghurt is less common but widespread. More than 300 types of bread are sold in bakery shops across the country.

As a country with many immigrants, Germany has adopted many international dishes into its cuisine and daily eating habits. Italian dishes like Pizza and Pasta, Turkish and Arab dishes like Doner Kebab and Falafel are well established, especially in bigger cities. International burger chains, as well as Chinese and Greek restaurants, are widespread. Indian, Thai, Japanese, and other Asian cuisines have gained popularity in recent decades. Among nine high-profile restaurants in Germany, the Michelin guide has awarded three stars, the highest designation, while 15 more received two stars. German restaurants have become the world’s second most decorated eateries after France. Although wine is becoming more popular in many parts of Germany, the national alcoholic drink is beer. German beer consumption per person is declining but—at 116 litres annually—it is still among the highest in the world. Beer varieties include Alt, Bock, Dunkel, Kolsch, Lager, Malzbier, Pils, and Weizenbier. Among 18 surveyed western countries, Germany ranked 14th in the list of per capita consumption of soft drinks in general, while it ranked third in the consumption of fruit juices. Furthermore, carbonated mineral water and *Schorle* (its mixture with fruit juice) are very popular in Germany.

Society

Since the 2006 World Cup celebrations the internal and external perception of Germany’s national image has changed. In annually conducted global surveys known as Nation Brands Index, Germany became significantly and repeatedly higher ranked after the tournament. People in 20 different states were asked to assess the country’s reputation in terms of culture, politics, exports, its people and its attractiveness to tourists, immigrants and investments. Germany has been named the world’s most valued nation among 50 countries in 2008. Another global opinion poll based on 29,977 responses in 28 countries for the BBC revealed that Germany is recognised for the most positive influence in the world in 2010, leading 28 investigated countries. A majority of 59% have a positive view of the country, while 14% have a negative view.

Germany is a legally and socially tolerant country towards homosexuals. Civil unions have been permitted since 2001. Gays and lesbians can legally adopt their partner's biological children (stepchild adoption). The Foreign Minister Guido Westerwelle and the mayors of the two largest German cities, Berlin and Hamburg, are openly gay.

During the last decade of the 20th century Germany has transformed its attitude towards immigrants considerably. Until the mid-nineties the opinion was widespread that Germany is not a country of immigration, even though about 10% of the population were of non-German origin. After the end of the influx of so-called *Gastarbeiter* (blue-collar guest-workers), refugees were a tolerated exception to this point of view. Many guest workers were of Turkish origins and brought an interesting influence into the country. Today the government and much of German society are acknowledging the opinion that controlled immigration should be allowed based on the qualification of immigrants.

With an expenditure of €67 billion on international travel in 2008, Germans spent more money on travel than any other country. The most visited foreign destinations were Spain, Italy and Austria.

Agriculture in Germany

Agriculture in Germany is a small sector of the German economy. It has declined in importance all during the twentieth century and by 1989 amounted to only 1.6 percent of West German GDP. Although agriculture's share of East German GDP was twice as high as in the west, even after the two economies are completely united, agriculture's share of GDP is expected to amount to only about 2 percent. However, despite the sector's small size, it remains politically important. The number of farms had decreased steadily in West Germany, from 1.6 million in 1950 to 630,000 in 1990. In East Germany, where farms were collectivized under the socialist regime, there had been about 5,100 agricultural production collectives with an average of 4,100 hectares under cultivation. Since unification, about three-quarters of the collectives have remained as cooperatives, partnerships, or joint-stock companies. The others were returned to their original owners—if those owners could be found—or were privately sold, becoming about 14,000 private farms. In western Germany and in the newly privatized farms in eastern Germany, family farms predominate. For the 630,000 farms, there are 750,000 full-time employees. There are also, however, many more part-time employees, and most farms do not represent their owners' full-time occupation.

Although the number of farms has declined, production has actually increased through more efficient production methods. By the early 1990s, a single farmer could produce enough food for seventy-five people, far more than was the case in the 1950s or 1960s. Agricultural products vary from region to region. Germany, known for the great population of 82. 5 million people, is consisted mostly of great lands of agriculture. In the flat terrain of northern Germany and especially in the eastern portions, cereals and sugar beets are grown. Elsewhere, with the terrain more hilly and even mountainous, farmers produce vegetables, milk, pork, or beef. Almost all large cities are surrounded by fruit orchards and vegetable farms. Most river valleys in southern and western Germany, especially along the Rhine and the Main, have vineyards. Beer is produced mainly, but not exclusively, in Bavaria.

Since the 1960s, German agricultural policy has not been made in Germany but in the EC. All agricultural laws and regulations are written in Brussels, often after difficult negotiations between food-producing and food-consuming states. The main objective of those negotiations is to obtain high incomes for the farmers while keeping market prices low enough to avoid consumer protests. To make up the difference, the EC adopted the Common Agricultural Policy subsidy program and the export subsidy program, both of which benefit German farmers as well as other EU farmers. In return, the German farmers have complied with European directives on the quality and quantity of production.

8

Norway

Norway, officially the Kingdom of Norway, is a Nordic country in Northern Europe occupying the western portion of the Scandinavian Peninsula, as well as Jan Mayen and the Arctic archipelago of Svalbard. Norway has a total area of 385,252 square kilometres (148,747 sq mi) and a population of about 4.8 million. It is one of the most sparsely populated countries in Europe. The majority of the country shares a border to the east with Sweden; its northernmost region is bordered by Finland to the south and Russia to the east; and Denmark lies south of its southern tip across the Skagerrak Strait. The capital city of Norway is Oslo. Norway's extensive coastline, facing the North Atlantic Ocean and the Barents Sea, is home to its famous fjords.

After World War II, Norway experienced rapid economic growth, with the first two decades due to the Norwegian shipping and merchant marine and domestic industrialization, and from the early 1970s, a result of exploiting large oil and natural gas deposits that had been discovered in the North Sea and the Norwegian Sea. Today, Norway ranks as the third wealthiest country in the world in monetary value, with the largest capital reserve per capita of any nation. Norway is the world's fifth largest oil exporter, and the petroleum industry accounts for around a quarter of its GDP. Following the ongoing financial crisis of 2007–2010, bankers have deemed the Norwegian krone to be one of the most solid currencies in the world.

Norway has rich resources of oil, natural gas, hydroelectric power, forests, and minerals, and was the second largest exporter of seafood (in value, after the People's Republic of China) in 2006. Other major industries include shipping, food processing, shipbuilding, the metal industry, chemicals, mining, fishing, and the pulp and paper products from forests. Norway maintains a Scandinavian welfare model with universal healthcare, subsidized higher education, and a comprehensive social security system. Norway was ranked highest of all countries in human development from 2001 to 2007, and then again in 2009. It was also rated the most peaceful country in the world in a 2007 survey by Global Peace Index.

Norway is a constitutional monarchy and parliamentary democracy, with King Harald V as its head of state and Jens Stoltenberg as its prime minister. It is a unitary state with administrative subdivisions on two levels known as counties (fylker) and municipalities (kommuner). The Sami people have a certain amount of self-determination and influence over traditional territories through the Sami Parliament and the Finnmark Act. Although having rejected European Union membership in two referendums, Norway maintains close ties with the union and its member countries, as well as with the United States. Norway remains one of the biggest financial contributors to the UN, and participates with UN forces in international missions, notably in Afghanistan, Kosovo, and Sudan. Norway is a founding member of the UN, NATO, the Council of Europe, and the Nordic Council, and is a member of the European Economic Area, the WTO, and the OECD.

Etymology

Norway is officially called *Kongeriket Norge* (The Kingdom of Norway) in the Bokmal written norm, and *Kongeriket Noreg* in the Nynorsk written norm.

The usual Old Norse form of *Norway* is *Noregr*, and the usual mediaeval Latin form *Nor(th)vegia*, though the earliest known written occurrence of the name is English (in the late-ninth-century account of the travels of Ohthere of Halogaland), in the form *noroweg*. Although some mediaeval texts attribute the name to a mythical King Norr, it is conventionally derived today from Old Norse **norovegr*, meaning "the northern route" (the way northwards). Norse also had the terms *austrvegr* "the lands in the east" (Russia and Asia), *vestrvegr* "the lands in the west" (the British Isles) and *suorvegr* "the lands in the south" (the Mediterranean). There is, however, some possibility that mediaeval forms in *noro-*, *north-*are folk-etymologizations and that the name has other origins.

History

Prehistory

Archaeological findings indicate the area currently constituting Norway has been inhabited since at least the 10th millennium BC. The indigenous people of Northern Norway and Central Norway are the Sami people, though Norse culture arrived very early also. The current monarch of Norway has stated that the kingdom was founded upon the territories of two peoples—the Norwegians and the Sami.

In the first centuries AD, Norway consisted of a number of petty kingdoms.

According to Jared Diamond, around AD 600 the Norse obtained the knowledge of sailing.

Viking Age

The Viking Age was characterized by expansion and emigration by Viking seafarers. According to tradition, Harald Fairhair (Harald Harfagre) unified them into one in 872 AD after the Battle of Hafrsfjord in Stavanger, thus becoming the first king of a united Norway. (The date of 872 may be somewhat arbitrary. In fact, the actual date may be just prior to 900 AD). Harald's realm was mainly a South Norwegian coastal state. Harald Fairhair ruled with a strong hand and, according to the sagas, many Norwegians left the country to live in Iceland, the Faroe Islands, Greenland, and parts of Britain and Ireland. The modern-day Irish cities of Limerick, Dublin, and Waterford were founded by Norwegian settlers.

Norse traditions were slowly replaced by Christianity in the 10th and 11th centuries. This is largely attributed to the missionary kings Olav Tryggvasson and St. Olav. Haakon the Good was Norway's first Christian king, in the mid tenth century, though his attempt to introduce the religion was rejected. Born sometime in between 963–969 AD, Olav Tryggvasson set off raiding in England with 390 ships. He attacked London during this raiding. Arriving back in Norway in 995, Olav landed in Moster. There he built a church which became the first Christian church ever built in Norway. From Moster, Olav sailed north to Trondheim were he has acclaimed King of Norway by the Eyrathing in 995 AD.

Feudalism never really developed in Norway and Sweden, as it did in the rest of Europe. However, the administration of government took on a very conservative feudal character. The Hanseatic League forced the royalty to cede to them greater and greater concessions over foreign trade and the economy. The League had this hold over the royalty because of the loans the Hansa had made to the royalty and the large debt the kings were carrying. The League's monopolistic control over the economy of Norway put pressure on all classes, especially the peasantry, to the degree that no real burgher class existed in Norway.

Kalmar Union

Upon the death of Haakon V, King of Norway, in 1319, three year-old Magnus Erickson inherited the throne as King Magnus VII of Norway. At the same time a movement to make Magnus King of Sweden proved successful. (At this time both the kings of Sweden and of Denmark were elected to the throne by their respective nobles.) Thus, with his election to the throne of Sweden, both Sweden and Norway were united under King Magnus VII.

In 1349, the Black Death radically altered Norway, killing between 50% and 60% of its population and leaving it in a period of social and economic decline. The plague left Norway very poor. Although the death rate was comparable with the rest of Europe, economic recovery took much longer because of the small thinly scattered population. Before the plague, the population was only about 500,000 people, and afterwards, many farms lay idle while the population slowly recovered. The few surviving farms' tenants found their bargaining positions with their landlords greatly strengthened.

King Magnus VII, noted above, ruled Norway until 1350, when his son, Haakon, was placed on the throne as Haakon VI. In 1363, Haakon VI married Margaret, the daughter of Danish King Valdemar. Upon the death of Haakon VI, in 1379, his son, Olaf V was only 10 years-old. Olaf had already been elected to the throne of Denmark on May 3, 1376. Thus, upon his ascension to the throne of Norway, Olaf united Denmark and Norway under a single throne. Olaf's mother and Haakon's widow, Queen Margaret, managed the foreign affairs of Denmark and Norway during the minority of Olaf V.

Margaret was working toward a union of Sweden with Denmark and Norway by having Olaf elected to the Swedish throne. She was on the verge of achieving this goal when Olaf V suddenly died. However, Denmark made Margaret temporary ruler upon the death of Olaf. On February 2, 1388, Norway followed suit and crowned Margaret.

Queen Margaret knew that her power would be more secure if she were able to find a king to rule in her place. She settled on Erik of Pomerania, grandson of her sister. Thus at an all-Scandinavian meeting held at Kalmar, Erik of Pomerania was crowned king of all three Scandinavian countries—four countries in actuality because Finland was part of Sweden at this time. Thus, royal politics resulted in personal unions between the Nordic countries, eventually bringing the thrones of Norway, Denmark, and Sweden under the control of Queen Margrethe I of Denmark when the country entered into the Kalmar Union.

Union with Denmark

After Sweden broke out of the Kalmar Union in 1521, Norway remained with Denmark until 1814, a total of 436 years. During the national romanticism of the 19th century, this period was by some referred to as the "400-Year Night", since all of the kingdom's royal, intellectual, and administrative power was centred in Copenhagen in Denmark.

With the introduction of Protestantism in 1536, the archbishopric in Trondheim was dissolved, and Norway effectually became a tributary to Denmark, and the church's incomes were distributed to the court in Copenhagen instead. Norway lost the steady stream of pilgrims to the relics of St. Olav at the Nidaros shrine, and with them, much of the contact with cultural and economic life in the rest of Europe. Additionally, Norway saw its land area decrease in the 17th century with the loss of the provinces Bahuslen, Jemtland, and Herjedalen to Sweden, as a result of numerous wars between Denmark–Norway and Sweden. To the north, however, its territory was increased by the acquisition of the northern provinces of Troms and Finnmark, at the expense of Sweden and Russia.

The famine of 1695–96 killed roughly 10% of Norway's population. At least nine severe harvest failures were recorded in the Scandinavian countries between 1740 and 1800, each resulting in a substantial rise of the death rate.

Union with Sweden (19th Century)

After Denmark–Norway was attacked by the United Kingdom, at the Battle of Copenhagen it entered into an alliance with Napoleon, with the war leading to dire conditions and mass starvation in 1812. As the Danish kingdom found itself on the losing side in 1814, it was forced, under terms of the Treaty of Kiel, to cede Norway to the king of Sweden, while the old Danish-Norwegian provinces of Iceland, Greenland and the Faroe Islands remained with the Danish crown.

Norway took this opportunity to declare independence, adopted a constitution based on American and French models, and elected the crown prince of Denmark-Norway Christian Frederik as king on 17 May 1814. This is the famous Syttende Mai (Seventeenth of May) holiday celebrated by Norwegians and Norwegian-Americans alike. *Syttende Mai* is also called Norwegian Constitution Day.

However, the decision to link Norway with Sweden caused the Norwegian-Swedish War to break out between Sweden and Norway, but as Sweden's military was not strong enough to defeat the Norwegian forces outright and Norway's treasury was not large enough to support a protracted war, and as British and Russian navies blockaded the Norwegian coast, both Norway and Sweden were forced to negotiate a settlement. Accordingly, on November 4, 1814, Norway was forced into entering the union with Sweden. Under this arrangement, Norway did, however, keep its liberal constitution and kept control of its own independent institutions, except for the foreign service. Following the Union with Sweden economic development of Norway remained slow.

This period also saw the rise of the Norwegian romantic nationalism, as Norwegians sought to define and express a distinct national character. The movement covered all branches of culture, including literature (Henrik Wergeland [1808–1845], Bjornstjerne Bjornson [1832–1910], Peter Christen Asbjornsen [1812–1845], Jorgen Moe [1813–1882]), painting, and even language policy, where attempts to define a native written language for Norway led to today's two official written forms for Norwegian: Bokmal and Nynorsk.

King Karl XIV Johan who came to the throne of Norway/ Sweden in 1818 was the first king following Norway's break from Denmark and the union with Sweden. Karl Johan was a complex man whose long reign extended to 1844. King Karl Johan protected the constitution and liberties of Norway/Sweden during the age of Metternich. As such he was regarded as a liberal monarch for that age. However, he was ruthless in his use of paid informers, the secret police and restrictions on the freedom of the press to put down public movements for reform—especially the Norwegian national independence movement.

The Romantic Era that followed the reign of King Karl Johan brought some significant social and political reforms. In 1854 women were given the right to inherit property in their own right just like men. In 1863, the last trace of keeping unmarried women in the status of minors was removed. Furthermore, women were then eligible for different occupations, particularly the common school teacher. However, by mid-century Norway was still far from a "democracy." Voting was limited to officials, property owners, leaseholders, and burghers of incorporated towns. There was some dissatisfaction with this backwardness.

Still Norway remained a conservative society. Life in Norway (especially economic life) was "dominated by the aristocracy of professional men who filled most of the important posts in the central government." There was no strong bourgeosie class in Norway to demand a breakdown of this aristocratic control of the economy. Thus, even while revolution swept over most of the countries of Europe in 1848, Norway was largely unaffected by revolts that year. Most revolts broke themselves on the granite conservativism of the Norwegian society. Indeed the Thrane movement was the only "revolt" that broke out in Norway in 1848.

Marcus Thrane was a utopian socialist. He made his appeal to the labouring classes urging a change of social structure "from below upwards." In 1848, he organized a labour society in Drammen. In just a few months this society had a membership of 500 and the society was publishing its own newspaper. Within two years 300 societies had been organized all over Norway with a total membership of 20,000 persons. The membership was drawn from the lower classes of both the town and country. For the first time these two groups felt they had common cause with each other. In the end, the revolt was easily crushed, Thrane was captured and sentenced to three years in jail for crimes against the safety of the state. Upon his release from jail, after serving his sentence, Marcus Thrane migrated to the United States of America.

Independence

Christian Michelsen, a shipping magnate and statesman, Prime Minister of Norway from 1905 to 1907, played a central role in the peaceful separation of Norway from Sweden on 7 June 1905. After a national referendum confirmed the people's preference for a monarchy over a republic, the Norwegian government offered the throne of Norway to the Danish Prince Carl, and Parliament unanimously elected him king, the first king of a fully independent Norway in 586 years. He took the name of Haakon VII, after the mediaeval kings of independent Norway.

In 1898, all men were granted universal suffrage, followed by all women in 1913.

World Wars I and II

During World War I, Norway was a neutral country. In reality, however, Norway had been pressured by Great Britain to hand over increasingly large parts of its massive merchant fleet to Britain at low rates, as well as to join the trade blockade against Germany. Norwegian merchant marines ship with Norwegian sailors were then required to sail under the British flag and risk being sunk by German submarines. Thus, many Norwegian sailors and ships were lost. Thereafter, the world ranking of the Norwegian merchant marine fell from fourth place in the world to sixth place in the world.

Norway also proclaimed its neutrality during World War II, but Norway was invaded by German forces on 9 April 1940. Norway was unprepared for the German surprise attack, so military and naval resistance only lasted for two months. The armed forces in the north launched an offensive against the German forces in the Battles of Narvik, until they were forced to surrender on June 10 after losing British help following the Fall of France.

King Haakon and the Norwegian government escaped to Rotherhithe, London, England, and they supported the fight through inspirational radio speeches from London and by supporting clandestine military actions in Norway against the Nazis. On the day of the invasion, the collaborative leader of the small National-Socialist party Nasjonal Samling, Vidkun Quisling, tried to seize power but was forced by the German occupiers to step aside. Real power was wielded by the leader of the German occupation authority, Reichskommissar Josef Terboven. Quisling, as *minister president*, later formed a collaborationist government under German control. Up to 15,000 Norwegians volunteered to fight in German units, including the Waffen-SS.

There were also many Norwegians, and those of Norwegian descent, that joined the allied forces as well as the Free Norwegian Forces. From the small group that had left Norway in June 1940 consisting of 13 ships, five aircraft and 500 men from the Royal Norwegian Navy who followed the King to the United Kingdom the force had grown by the end of the war to 58 ships and 7,500 men in service in the Norwegian Navy; 5 Squadrons of aircraft (including Spitfires, Sunderland flying boats and Mosquitos) in the newly formed Norwegian Air Force; and land forces including the Norwegian Independent Company 1 & 5 Troop as well as No.10 Commandos.

During the five years of Nazi occupation, Norwegians built a resistance movement which fought the German occupation forces with both civil disobedience and armed resistance including the destruction of Norsk Hydro's heavy water plant and stockpile of heavy water at Vemork, which crippled the German nuclear program. More important to the Allied war effort, however, was the role of the Norwegian Merchant Marine. At the time of the invasion, Norway had the fourth largest merchant marine fleet in the world. It was led by the Norwegian shipping company Nortraship under the Allies throughout the war and took part in every war operation from the evacuation of Dunkirk to the Normandy landings. Each December Norway gives a Christmas tree to the United Kingdom as thanks for the British assistance during World War II. A ceremony takes place to erect the tree in London's famous Trafalgar Square.

Post-war History

From 1945 to 1961, the Labour Party held an absolute majority in the parliament. The government, led by prime minister Einar Gerhardsen, embarked on a program inspired by Keynesian economics, emphasizing state financed industrialization, cooperation between trade unions and employers' organizations. Many measures of state control of the economy imposed during the war were continued, although the rationing of dairy products was lifted in 1949, while price control and rationing of housing and cars continued as long as until 1960.

The wartime alliance with Great Britain and the United States was continued in the post-war years. Although pursuing the goal of a socialist economy, the Labour Party distanced itself from the communists (especially after Soviet seizure of power in Czechoslovakia in 1948), and strengthened its foreign policy and defence policy ties with the U.S. Norway received Marshall Plan aid from the United States starting in 1947, joined the OEEC one year later and became a founding member of the North Atlantic Treaty Organization (NATO) in 1949.

Around 1975, both the proportion and absolute number of workers in industry peaked. Since then labour intensive industries and services like factory mass production and shipping have largely been outsourced.

In 1969, the Phillips Petroleum Company discovered petroleum resources at the Ekofisk field west of Norway. In 1973, the Norwegian government founded the State oil company, Statoil. Oil production did not provide net income until the early 1980s because of the large capital investment that was required to establish the country's petroleum industry.

Norway was a founding member of the European Free Trade Area (EFTA). Two referendums on joining the European Union failed by narrow margins in 1972 and 1994. In 1981, a Conservative government led by Kare Willoch replaced the Labour Party with a policy of stimulating the stagflated economy with tax cuts, economic liberalization, deregulation of markets, and measures to curb the record-high inflation (13.6% in 1981).

Norway's first female prime minister, Gro Harlem Brundtland of the Labour party, continued many of the reforms of her right-wing predecessor, while backing traditional Labour concerns such as social security, high taxes, the industrialization of nature, and feminism. By the late 1990s, Norway had paid off its foreign debt and had started accumulating a sovereign wealth fund. Since the 1990s, a divisive question in politics has been how much of the income from petroleum production the government should spend, and how much it should save.

Geography, Climate, and Environment

Norway comprises the western part of Scandinavia in Northern Europe. The rugged coastline, broken by huge fjords and thousands of islands, stretches 2,500 kilometres (1,600 mi) and 83,000 kilometres (52,000 mi) including fjords and islands. Norway shares a 1,619-kilometre (1,006 mi) land border with Sweden, 727 kilometres (452 mi) with Finland and 196 kilometres (122 mi) with Russia at the east. To the north, west and south, Norway is bordered by the Barents Sea, the Norwegian Sea, the North Sea and Skagerak.

At 385,252 square kilometres (148,747 sq mi) (including Svalbard and Jan Mayen), much of the country is dominated by mountainous or high terrain, with a great variety of natural features caused by prehistoric glaciers and varied topography. The most noticeable of these are the fjords: deep grooves cut into the land flooded by the sea following the end of the Ice Age. The longest is Sognefjorden at 204 kilometres (127 mi). Sognefjorden is the world's second deepest fjord, and Hornindalsvatnet is the deepest lake in Europe. Frozen ground all year can be found in the higher mountain areas and in the interior of Finnmark county. Numerous glaciers are found in Norway.

The land is mostly made of hard granite and gneiss rock, but slate, sandstone and limestone are also common, and the lowest elevations contain marine deposits. Because of the Gulf Stream and prevailing westerlies, Norway experiences higher temperatures and more precipitation than expected at such northern latitudes, especially along the coast. The mainland experiences four distinct seasons, with colder winters and less precipitation inland. The northernmost part has a mostly maritime Subarctic climate, while Svalbard has an Arctic tundra climate.

The southern and western parts of Norway experience more precipitation and have milder winters than the southeastern part. The lowlands around Oslo have the warmest and sunniest summers but also cold weather and snow in wintertime (especially inland). Average temperatures have risen the last decades, decreasing the amount of days with snow cover in the lowlands.

Because of the large latitudinal range of the country and the varied topography and climate, Norway has a larger number of different habitats than almost any other European country. There are approximately 60,000 species in Norway and adjacent waters (excluding bacteria and virus). The Norwegian Shelf large marine ecosystem is considered highly productive.

The total number of species include 16,000 species of insects (probably 4,000 more species yet to be described), 20,000 species of algae, 1,800 species of lichen, 1,050 species of mosses, 2,800 species of vascular plants, up to 7,000 species of fungi, 450 species of birds (250 species nesting in Norway), 90 species of mammals, 45 fresh-water species of fish, 150 salt-water species of fish, 1,000 species of fresh-water invertebrates and 3,500 species of salt-water invertebrates. About 40,000 of these species have been described by science. The red list of 2006 encompasses 3,886 species.

Seventeen species are listed mainly because they are endangered on a global scale, such as the European Beaver, even if the population in Norway is not seen as endangered. There are 430 species of fungi on the red list, many of these are closely associated with the small remaining areas of old-growth forests. There are also 90 species of birds on the list and 25 species of mammals. 1,988 current species are listed as endangered or vulnerable as of 2006; of these are 939 listed as vulnerable (VU), 734 species are listed as endangered (EN), and 285 species are listed as critically endangered (CR) in Norway, among these are the grey wolf, the arctic fox (healthy population on Svalbard) and the pool frog. The largest predator in Norwegian waters is the sperm whale, and the largest fish is the basking shark. The largest predator on land is the polar bear, while the brown bear is the largest predator on the Norwegian mainland, where the common moose is the largest animal.

Because of Norway's high latitude, there are large seasonal variations in daylight. From late May to late July, the sun never completely descends beneath the horizon in areas north of the Arctic Circle (hence Norway's description as the "Land of the Midnight Sun"), and the rest of the country experiences up to 20 hours of daylight per day. Conversely, from late November to late January, the sun never rises above the horizon in the north, and daylight hours are very short in the rest of the country.

Throughout Norway, one will find stunning and dramatic scenery and landscape. The west coast of southern Norway and the coast of northern Norway present some of the most visually impressive coastal sceneries in the world. National Geographic has listed the Norwegian fjords as the world's top tourist attraction. The 2008 Environmental Performance Index put Norway in second place, after Switzerland, based on the environmental performance of the country's policies.

Government and Politics

According to the Constitution of Norway, which was adopted on 17 May 1814 and inspired by the United States Declaration of Independence and French revolution of 1776 and 1798, respectively, Norway is a unitary constitutional monarchy with a parliamentary system of government, wherein the King of Norway is the head of state and the Prime Minister is the head of government. Power is separated between the legislative, executive and judicial branches of government, as defined by the Constitution, which serves as the country's supreme legal document.

The Monarch officially retains executive power, however, following the introduction of a parliamentary system of government, the duties of the Monarch have since become strictly representative and ceremonial, such as the formal appointment and dismissal of the Prime Minister and other ministers in the executive government. Accordingly, the Monarch is commander-in-chief of the Norwegian armed forces, supreme authority in the Church of Norway, and serves as chief diplomatic representative abroad and a symbol of unity.

In practice, it is the Prime Minister who is responsible for the exercise of executive powers. Since his accession in 1991, Harald V of the House of Schleswig-Holstein-Sonderburg-Glucksburg has been King of Norway, the first in many years who has actually been born in the country. Haakon, Crown Prince of Norway is the legal and rightful heir to the throne and the Kingdom.

Constitutionally, legislative power is vested with both the government and the Parliament of Norway, but the latter is the supreme legislature and a unicameral body. A proposition can become a law or an act by simple majority amongst the 150 representatives, who are elected on the basis of proportional representation from 19 constituencies for four-year terms. An additional 19 seats ("levelling seats") are allocated on a nationwide basis to make the representation in parliament correspond better with the popular vote.

As a result, there are currently 169 Members of Parliament altogether. There is also a 4% election threshold to gain levelling seats in Parliament. As such, Norway is fundamentally structured as a representative democracy. Effectively called the *Storting*, meaning *Grand Assembly*, members of Parliament ratify treaties and can impeach members of the government if their acts are declared unconstitutional, and as such have the power to remove them from office in case of an impeachment trial.

The position of Prime Minister, Norway's head of government, is allocated to the Member of Parliament who can obtain the confidence of a majority in Parliament, usually the current leader of the largest political party or more effectively through a coalition of parties, as a single party normally doesn't have the support to form a government on its own. However, Norway has often been ruled by minority governments.

The Prime Minister nominates the Cabinet, traditionally drawn from members of the same political party in the Storting, to which they are responsible, and as such forms the executive government and exercises power vested to them by the Constitution. In order to form a government, however, more than half the membership of the Cabinet is required to belong to the Church of Norway. Currently, this means at least ten out of the 19 ministries. This has sparked controversy regarding an ongoing debate of separation of church and state in Norway. The current Prime Minister is Jens Stoltenberg, the leader of the Norwegian Labour Party (AP).

Through the Council of State, a privy council presided over by the Monarch, the Prime Minister and the Cabinet meet at the Royal Palace and formally consult the Monarch. Besides enacting parliamentary bills, all government bills need the formal approval by the Monarch before and after introduction to Parliament. Approval is also given by the Council to all of the Monarch's actions as head of state. Although all government and parliamentary acts are decided beforehand, the privy council is an example of another symbolic gesture the King obtains.

Members of the Storting are directly elected from party-lists proportional representation in nineteen plural-member constituencies in a national multi-party system. Historically, both the Norwegian Labour Party and Conservative Party have played leading political roles, while the former has remained in power since the 2005 election, in a Red-Green coalition with the Socialist Left Party and the Centre Party.

Since then, both the Conservative Party and the Progress Party have won great amount of seats in the Parliament, however, as of the 2009 general election, not sufficient enough to overthrow the coalition. This has been the result of poor cooperation with the Liberal Party and the Christian Democratic Party. As such, Jens Stoltenberg, the leader of the Labour Party, remains Prime Minister of Norway with the necessary majority attributed to the alliance with the Socialist Left and Centre parties.

Judicial System and Law Enforcement

Norway uses a civil law system where laws are created and amended in Parliament and the system regulated through the Courts of Justice of Norway. It consists of the Supreme Court of 19 permanent judges and a Chief Justice, appellate courts, city and district courts, and conciliation councils. The judiciary, although traditionally a third branch of government, is independent of executive and legislative branches. While the Prime Minister nominates Supreme Court Justices for office, their nomination must be approved by Parliament and formally confirmed by the Monarch in the Council of State. Usually, judges attached to regular courts are formally appointed by the Monarch on the advice of the Prime Minister.

The Courts' strict and formal mission is to regulate the Norwegian judicial system, interpret the Constitution, and as such implement the legislation adopted by Parliament and monitor the legislative and executive powers to ensure that they themselves comply with the acts of legislation that have been previously adopted.

Law enforcement in Norway is carried out by the Norwegian Police Service. The Norwegian Police Service is a Unified National Police Service made up of 27 Police Districts and several specialist agencies like the Norwegian National Authority for the Investigation and Prosecution of Economic and Environmental Crime and Kripos, each headed by a Chief of Police. The Police Service is headed by the National Police Directorate, which in turn is subordinate to the Ministry of Justice and the Police, the Police Directorate is headed by a National Police Commissioner. The only exception is the Norwegian Police Security Agency who answers directly to the Ministry of Justice and the Police.

In its 2007 Worldwide Press Freedom Index, Reporters Without Borders ranked Norway at a shared 1st place (with Iceland) out of 169 countries. The death penalty was abolished in Norway in 1902. Death penalty for high treason in war and war-crimes was also abolished in 1979. Currently, Norway has the lowest homicide rate in the world.

Foreign Relations and Military

Norway maintains embassies in 86 countries. 60 countries maintain an embassy in Norway, all of them in the capital, Oslo.

Norway is a founding member of the United Nations, NATO and the Council of Europe. The Norwegian electorate has twice rejected treaties of accession to the European Union (EU), although most legislation made by the EU is implemented in the country because of Norway's membership in the European Economic Area (EEA). This ensures Norway's access to the EU's internal market. Norway has been considered a prominent participant in international development, having been heavily involved diplomatically with the failed Oslo Accords regarding the longtime conflict between Israel and Palestinians in the Middle East. Norway maintains close diplomatic relations with the United States.

The Norwegian Armed Forces currently numbers about 23,000 personnel, including civilian employees. According to the current (as of 2009) mobilization plans, the strength during full mobilization is approximately 83,000 combatant personnel. Norway has conscription for males (6–12 months of training) and voluntary service for females. The Armed Forces are subordinate to the Norwegian Ministry of Defence and the Commander-in-Chief is King Harald V. The military of Norway is divided into the following branches: the Army, the Royal Navy, the Royal Air Force and the Home Guard.

Partly due to Norway's failure to maintain its traditional policy of neutrality in World War II (surrendered to Nazi Germany in June 1940), the country was one of the founding nations of the North Atlantic Treaty Organization (NATO) on 4 April 1949. At present, Norway contributes in the International Security Assistance Force (ISAF) in Afghanistan. Notable international missions in recent years with Norwegian participation include the following:

- Kosovo: Kosovo Force (KFOR) and United Nations Interim Administration Mission in Kosovo (UNMIK)
- Afghanistan: International Security Assistance Force (ISAF)
- Bosnia: (in NATO/EUFOR HQ and Liaison Observer Team in Cazin)
- Sudan: United Nations Mission in Sudan (UNMIS)
- Lebanon: United Nations Interim Force in Lebanon (UNIFIL).

Administrative Divisions

Norway, a unitary state, is effectively divided into five regions, although this is for strategic geographical purposes only. The regions do not have their own administrative form of local government, nor a directly elected assembly. The regions are further divided into nineteen first-level administrative counties (*fylker*). The counties are administrated through directly elected county assemblies who elect the County Governor. Additionally, the King and government are represented in every county by a fylkesmann, who effectively acts as a Governor. As such, the Government is directly represented at a local level through the County Governors' offices. The counties are then sub-divided into 430 second-level municipalities (*kommuner*), which in turn are administrated by directly elected municipal council, headed by a mayor and a small executive cabinet. The capital of Oslo is considered both a county and a municipality. Norway also has two integral overseas territories, Jan Mayen and Svalbard. There are three Antarctic and Subantarctic dependencies: Bouvet Island, Peter I Island and Queen Maud Land.

In addition, there are 96 settlements with city status in Norway. In most cases, the city borders are coterminous with the borders of their respective municipalities. Often, Norwegian city municipalities include large non-built up areas; for example, Oslo municipality contains large forests, located north and southeast of the city, and over half of Bergen municipality consists of mountainous areas.

Economy

The economy of Norway is a developed mixed economy with heavy state-ownership in strategic areas of the economy. Although sensitive to global business cycles, the economy of Norway has shown robust growth since the start of the industrial era. Shipping has long been a support of Norway's export sector, but much of Norway's economic growth has been fueled by an abundance of natural resources, including petroleum exploration and production, hydroelectric power, and fisheries. Agriculture and traditional heavy manufacturing have suffered relative decline compared to services and oil-related industries, and the public sector is among the largest in the world as a percentage of the overall gross domestic product. The country has a very high standard of living compared with other European countries, and a strongly integrated welfare system.

Pre-industrial Revolution

Prior to the industrial revolution, Norway's economy was largely based on agriculture, timber, and fishing. Norwegians typically lived under conditions of considerable scarcity, though famine was rare. Except for certain fertile areas in Hedemarken and Ostfold, crops were limited to hardy grains, such as oats, rye, and barley; and livestock to sheep, goats, cattle, pigs, and some poultry; in places this was complemented with hunting.

In areas of Central and Northern Norway, the Sami subsisted on the nomadic herding of reindeer. Fishing all around the coast was dangerous work, though fish such as herring, cod, halibut, and other cold-water species were found in abundance. The introduction of the potato to Norway provided considerable relief for Norwegians.

All around the coast, the harvesting of fish (including cod, herring, halibut, and other cold water species) was an important supplement to farming and was in many areas in the north and west the primary household subsistence. Fishing was typically supplemented with crop-growing and the raising of livestock on small farms.

The economic conditions in Norway did not lend themselves to the formation of feudal system, though several kings did reward land to loyal subjects who became knights. Self-owning farmers were—and continue to be—the main unit of work in Norwegian agriculture, but leading up to the 19th century farmers ran out of land available for farming. Many agricultural families were reduced to poverty as tenant farmers, and served as the impetus for emigration to North America.

Merchant and civil servant classes were small in Norway. In some cities, notably Bergen, trading centres grew up around import and export of various goods, especially stockfish, going back to the Viking era. Centres for shipping grew along the coast, and a small number of shipping magnates were among the first truly wealthy people in Norway.

Industrial Revolution

Aside from mining in Kongsberg and Roros, industrialization came with the first textile mills that were built in Norway in the middle of the 19th century. But the first large industrial enterprises came into formation when entrepreneurs politics, leading to the founding of banks to serve those needs.

Industries also offered employment for a large number of individuals who were displaced from the agricultural sector. As wages from industry exceeded those from agriculture, the shift started a long-term trend of reduction in cultivated land and rural population patterns. The working class became a distinct phenomenon in Norway, with its own neighbourhoods, culture, and politics.

Social Democratic Reforms and State Ownership

The roots of the socialist movement in Norway were based on dangerous working conditions, exploitative labour relations policies, and the demand for collective bargaining. As socialism became part of the mainstream labour movement, it also became part of the mainstream political discourse.

The state has large ownership positions in key industrial sectors, such as the strategic petroleum sector (Statoil), hydroelectric energy production (Statkraft), aluminum production (Norsk Hydro), the largest Norwegian bank (DnB NOR) and telecommunication provider (Telenor). The government controls 31.6% of publicly-listed companies. When non-listed companies are included the state has even higher share in ownership (mainly from direct oil license ownership).

After World War II, the Norwegian Labour Party, with Einar Gerhardsen as prime minister, embarked on a number of social democratic reforms aimed at flattening the income distribution, eliminating poverty, ensuring social services such as retirement, medical care, and disability benefits to all, and putting more of the capital into the public trust.

As a result, the public sector grew as a percentage of the overall economy. Highly progressive income taxes, the introduction of value-added tax, and a wide variety of special surcharges and taxes made Norway one of the most heavily taxed economies in the world. Authorities particularly taxed discretionary spending, levying special taxes on automobiles, tobacco, alcohol, cosmetic items, and so on.

Norway's long-term social democratic policies, extensive governmental tracking of information, and the homogeneity of its population lent themselves particularly well for economic study, and academic research from Norway proved to make significant contributions to the field of macroeconomics during this era. When Norway became a petroleum-exporting country, the economic effects came under further study.

Petroleum and Post-industrialism

Oil-exporting Country

In May of 1963, Norway asserted sovereign rights over natural resources in its sector of the North Sea. Exploration started on July 19, 1966, when Ocean Traveler drilled its first well. Initial exploration was fruitless, until Ocean Viking found oil on August 21, 1969. By the end of 1969, it was clear that there were large oil and gas reserves in the North Sea. The first oil field was Ekofisk, produced 427,442 barrels of crude in 1980. Since then, large natural gas reserves have also been discovered.

Against the backdrop of the Norwegian referendum not to join the European Union, the Norwegian Ministry of Industry, headed by Ola Skjak Break moved quickly to establish a national energy policy. Norway decided to stay out of OPEC, keep its own energy prices in line with world markets, and spend the revenue-known as the "currency gift"-wisely. The Norwegian government established its own oil company, Statoil, and awarded drilling and production rights to Norsk Hydro and the newly formed Saga Petroleum.

The North Sea turned out to present many technological challenges for production and exploration, and Norwegian companies invested in building capabilities to meet these challenges. A number of engineering and construction companies emerged from the remnants of the largely lost shipbuilding industry, creating centres of competence in Stavanger and the western suburbs of Oslo. Stavanger also became the land-based staging area for the offshore drilling industry. Presently North Sea is past its peak oil production.

Reservations about European Union

On September 24 and 25 1972, the Norwegian parliament put to a referendum the question whether Norway should join the European Union. The proposal was turned down with a slim margin. The Norwegian government proceeded to negotiate a trade agreement with the EU that would give Norwegian companies access to European markets. Over time, Norway renegotiated and refined this agreement, ultimately joining the European Free Trade Association and the European Economic Area.

Although Norway's trade policies have long aimed at harmonizing its industrial and trade policy with the EU's, a new referendum in 1994 gave the same result as in 1972, and Norway remains one of only two Nordic countries outside the EU, the other being Iceland.

Although much of the highly divisive public debate about EU membership turned on political rather than economic issues, it formed economic policy in several important ways:

- Both politicians and the public came to terms with the fact that Norway's economic development was dependent on taking advantage of its comparative advantage by specializing in certain areas for export and relying on import for everything else. This has had a significant effect on Norway's agricultural policy, which has been reshaped to address population patterns rather than self-sufficiency.
- The proceeds from oil revenue could not fuel private or public consumption if Norway were to sustain its prosperity when oil reserves run out.
- In order to participate in European markets, Norway has had to open its domestic markets to European imports. Although some pricing and distribution issues (e.g., alcohol and automobiles) remain unresolved, Norway's consumer, capital, and employment markets are increasingly approaching those of Europe in general.

Norwegians have sought accommodations on a range of specific issues, such as products from fish farms, agricultural products, emission standards, etc., but these do not appear to differ substantially from those sought by bona fide EU members. It is expected that the issue of membership will be brought to a referendum again at some point.

Post-industrial Economic Developments

Several issues have dominated the debate on Norway's economy since the 1970s:

- Cost of living. Norway is among the most expensive countries in the world, as reflected in the Big Mac Index and other indices. Historically, transportation costs and barriers to free trade had caused the disparity, but in recent years, Norwegian policy in labour relations, taxation, and other areas have contributed significantly.

- Competitiveness of "mainland" industries. The high cost of labour and other structural features of the Norwegian environment have caused concern about Norway's ability to maintain its cost of living in a post-petroleum era. There is a clear trend toward ending the practice of "protecting" certain industries (*vernede industrier*) and making more of them "exposed to competition" (*konkurranseutsettelse*). In addition to interest in information technology, a number of small-to medium-sized companies have been formed to develop and market highly specialized technology solutions.
- The role of the public sector. The ideological divide between socialist and non-socialist views on public ownership has decreased over time. The Norwegian government has sought to reduce its ownership over companies that require access to private capital markets, and there is an increasing emphasis on government facilitating entrepreneurship rather than controlling (or restricting) capital formation. A residual distrust of the "profit motive" persists, and Norwegian companies are heavily regulated, especially with respect to labour relations.
- The future of the welfare state. Since World War II, successive Norwegian governments have sought to broaden and extend public benefits to its citizens, in the form of sickness and disability benefits, minimum guaranteed pensions, heavily subsidized or free universal health care, unemployment insurance, and so on. Public policy still favours the provision of such benefits, but there is increasing debate on making them more equitable and needs-based.
- Urbanization. For several decades, agricultural policy in Norway was based on the premise of minimal self-sufficiency. In later years, this has given way to a greater emphasis on maintaining population patterns outside of major urban areas. The term "district policy" (*distriktspolitikk*) has come to mean the demand that old and largely rural Norway is allowed to persist, ideally by providing them with a sustainable economic basis.

- Taxation. The primary purpose of the Norwegian tax system has been to raise revenue for public expenditures; but it is also viewed as a means to achieve social objectives, such as redistribution of income, reduction in alcohol and tobacco consumption, and as a disincentive against certain behaviours. Three elements of the tax system seem to attract the most debate:
  - Progressive taxation. At one time one of the most aggressive in the world, the top marginal tax rate on income has been decreased over time. In addition, Norwegians are taxed for their stated net worth, which some have argued discourages savings.
  - Value-added tax. The largest source of government revenue. The current standard rate is 25%, food and drink is 14%, and movie theater tickets and public transportation 7%.
  - Special surcharges and taxes. The government has established a number of taxes related to specific purchases, including cars, alcohol, tobacco, and various kinds of benefits.
  - Svalbard. People living on Svalbard (Spitsbergen) does not pay any tax at all, nor is it any value-added tax. This is mainly because the government wants the island to be more attractive to live on. Although food is more expensive here than at the main land-Mostly caused by the shipping costs, other things like vehicles, tobacco and alcohol are significantly cheaper because of the non-existing taxes. For example, a pack of tobacco that costs around 170 NOK on the mainland, costs around 25-30 NOK here.
- Environmental concerns. A number of political issues have had their origins in economic concerns, including the refineries at Mongstad and the hydroelectric power plant at Alta.

Economic Structure and Sustained Growth

The emergence of Norway as an oil-exporting country has raised a number of issues for Norwegian economic policy. There has been concern that much of Norway's human capital investment has been concentrated in petroleum-related industries. Critics have pointed out that Norway's economic structure is highly dependent on natural resources that do not require skilled labour, making economic growth highly vulnerable to fluctuations in the demand and pricing for these natural resources. The Government Pension Fund of Norway is part of several efforts to hedge against dependence on petroleum revenue.

Because of the oil boom since the 1970s, there has been little extensive government incentive to help develop and encourage new industries in the private sector, in contrast to other Nordic countries like Sweden and particularly Finland. However the last decades have started to see some incentive on national and local government levels to encourage formation of new "mainland" industries that are competitive internationally. In addition to aspirations for a high-tech industry, there is growing interest in encouraging small business growth as a source of employment for the future. In 2006, the Norwegian government formed nine "centres of expertise" to facilitate this business growth. Later in June 2007, the government contributed to the formation of the Oslo Cancer Cluster (OCC) as a center of expertise, capitalizing on the fact that 80% of cancer research in Norway takes place in proximity to Oslo and that most Norwegian biotechnology companies are focused on cancer.

There is continuing debate over the role of the public sector in Norway's economic development. Although there is broad consensus that Norway should pursue a mixed economic model, there is a persistent ideological schism between those who favour free market forces vs. socialist mechanisms.

In 2007, Norway saw a massive 6% growth of its economy, out pacing any other western nation. However, the growth was mostly due to increased consumer demand, and is expected to slow down in 2008.

Norwegians enjoy the second highest GDP per-capita (after Luxembourg) and third highest GDP (PPP) per-capita in the world. Norway maintained first place in the world in the UNDP Human Development Index (HDI) for six consecutive years (2001–2006), and then reclaimed this position in 2009.

The Norwegian economy is an example of a mixed economy, a prosperous capitalist welfare state featuring a combination of free market activity and large state ownership in certain key sectors. The state has large ownership positions in key industrial sectors, such as the strategic petroleum sector (Statoil), hydroelectric energy production (Statkraft), aluminium production (Norsk Hydro), the largest Norwegian bank (DnB NOR), and telecommunication provider (Telenor). Through these big companies, the government controls approximately 30% of the stock values at the Oslo Stock Exchange. When non-listed companies are included, the state has even higher share in ownership (mainly from direct oil license ownership). Norway is a major shipping nation and has the world's 6th largest merchant fleet, with 1,412 Norwegian-owned merchant vessels.

Referendums in 1972 and 1994 indicated that the Norwegian people wished to remain outside the European Union (EU). However, Norway, together with Iceland and Liechtenstein, participates in the European Union's single market via the European Economic Area (EEA) agreement. The EEA Treaty between the European Union countries and the EFTA countries– transposed into Norwegian law via "EOS-loven"– describes the procedures for implementing European Union rules in Norway and the other EFTA countries. This makes Norway a highly integrated member of most sectors of the EU internal market. However, some sectors, such as agriculture, oil and fish, are not wholly covered by the EEA Treaty. Norway has also acceded to the Schengen Agreement and several other intergovernmental agreements between the EU member states.

The country is richly endowed with natural resources including petroleum, hydropower, fish, forests, and minerals. Large reserves of petroleum and natural gas were discovered in the 1960s, which led to a boom in the economy. Norway has obtained one of the highest standards of living in the world in part by having a large amount of natural resources compared to the size of the population. The Norwegian welfare state makes public health care free, and parents have 12 months paid parental leave. The income that the state receives from natural resources includes a significant contribution from petroleum production and the substantial and well-managed income related to this sector. Norway has a very low unemployment rate, currently 3.1%. The hourly productivity levels, as well as average hourly wages in Norway are among the highest in the world. The egalitarian values of the Norwegian society ensure that the wage difference between the lowest paid worker and the CEO of most companies is much smaller than in comparable western economies. This is also evident in Norway's low Gini coefficient.

Ccst of living is about 30% higher in Norway than in the United States and 25% higher than the United Kingdom. The standard of living in Norway is among the highest in the world. Foreign Policy Magazine ranks Norway last in its Failed States Index for 2009, judging Norway to be the world's most well-functioning and stable country. Continued oil and gas exports coupled with a healthy economy and substantial accumulated wealth lead to a conclusion that Norway will remain among the richest countries in the world in the foreseeable future.

Resources

Export revenues from oil and gas have risen to 45% of total exports and constitute more than 20% of the GDP. Norway is the fifth largest oil exporter and third largest gas exporter in the world, but it is not a member of the OPEC. To reduce over-heating in economy from oil revenues and minimize uncertainty from volatility in oil price, and to provide cushion for the effect of aging of the population, the Norwegian government in 1995 established the sovereign wealth fund ("Government Pension Fund — Global"), which would be funded with oil revenues, including taxes, dividends, sales revenues and licensing fees.

The government controls its petroleum resources through a combination of state ownership in major operators in the oil fields (with approximately 62% ownership in Statoil in 2007) and the fully state-owned Petoro, which has a market value of about twice Statoil, and SDFI. Finally, the government controls licensing of exploration and production of fields. The fund invests in developed financial markets outside Norway. The budgetary rule ("Handlingsregelen") is to spend no more than 4% of the fund each year (assumed to be the normal yield from the fund).

By January 2006, the Government Pension Fund of Norway controlled assets valued at US$200 billion. During the first half of 2007, the pension fund became the largest fund in Europe, with assets of about US$300 billion (equivalent to over US$62,000 per capita). The savings equal the Norwegian GDP and are the largest capital reserve per capita of any nation as of April 2007. Projections indicate that the Norwegian pension fund may become the largest capital fund in the world. Currently it is the second-largest state-owned sovereign wealth fund, second only to the Abu Dhabi Investment Authority; Conservative estimates tell that the fund may reach US$800–900 billion by 2017. As of November 2009, the size of the fund is approximately US$455 billion, and it controls approximately 1.25% of all listed shares in Europe and more than 1% of the all the publicly traded shares in the world. The Norwegian Central Bank operates investment offices in London, New York and Shanghai. New guidelines (implemented in 2007) allow the fund to invest up to 60% of the capital in shares (maximum of 40% prior), while the rest may be placed in bonds and real-estate. As the stock markets tumbled in September 2008, the fund was able to buy more shares at low prices. In this way, the losses incurred by the market turmoil was recuperated by November 2009.

Other natural resource-based economies, such as Russia, are trying to learn from Norway by establishing similar funds. The investment choices of the Norwegian fund are directed by ethical guidelines; for example, the fund is not allowed to invest in companies that produce parts for nuclear weapons. The highly transparent investment scheme is lauded by the international community.

The future size of the fund is of course closely linked to the price of oil and to developments in international financial markets. The Norwegian trade surplus for 2008 reached approximately US$80 billion. With an enormous amount of cash invested in international financial markets, Norway has financial muscles to avert many of the worst effects of the financial crisis that hit most countries in the fall of 2008. As most western countries struggle with burgeoning foreign debt, Norway remains an island of stowed-away wealth, financial stability and economic power to meet the challenges of the worldwide economic crisis. In spite of the crisis, Norway still runs a 9% state budget surplus, being the only western country to run a surplus as of July 2009. In 2000, the government sold one-third of the state-owned oil company Statoil in an IPO. The next year, the main Telecom supplier, Telenor, was listed on Oslo Stock Exchange. The state also owns significant shares of Norway's largest bank, DnB NOR and the airline SAS. Since 2000, economic growth has been rapid, pushing unemployment down to levels not seen since the early 1980s (unemployment in 2007: 1.3%). The international financial crisis has primarily affected the industrial sector, but it is unlikely that unemployment will surpass 3,5% in 2009–2010. Norway is among the least affected countries of the international economic downturn. Neighbouring Sweden is experiencing substantially higher actual and projected unemployment numbers as a result of the ongoing recession, and in the 1st quarter of 2009 the GNP of Norway surpassed Sweden's for the first time in history, despite a population numbering about half of Sweden's.

Norway is also the world's second largest exporter of fish (in value, after China). Hydroelectric plants generate roughly 98–99% of Norway's electric power.

Transport

Transport in Norway is highly influenced by Norway's low population density, narrow shape and long coastline. Norway has old water transport traditions, but rail, road and air transport have increased in importance during the 20th century. Due to the low population density, public transport is less built out than in many European countries, especially outside the cities.

The main governing body is the Norwegian Ministry of Transport and Communications, which performs operations through numerous subsidiaries. Tasks related to public transport and some roads have been delegated to the counties and municipalities. Most infrastructure is publicly owned, while most operations are performed by private companies; public transport is subsidized.

In average each Norwegian transported themselves for 70 minutes each day. 8% of passenger transport was made by public transport; road transport is the dominant mode of transport. The transport sector was responsible for 4.1% of the gross national product and 6.6% of employment in 2006.

Aviation

Aviation has become an important passenger transport mode since the 1960s. Aircraft are a common used mode of transport on longer distances, and the routes between Oslo and Bergen, Trondheim and Stavanger are all among the ten largest in Europe. In Western and Northern Norway, with difficult terrain and lack of rail transport, regional airline travel provides quick travel within the region or to the capital.

Airports

Of the 98 airports in Norway, 51 are public, and 46 are operated by the state-owned Avinor. Seven airports have more than one million passengers annually. 41,089,675 passengers passed through Norwegian airports in 2007, of which 13,397,458 were international.

The main gateway by air to Norway is Oslo Airport, Gardermoen, located about 50 kilometres (31 mi) north of Oslo with departures to most European countries and some intercontinental destinations. It is hub for the two major Norwegian airlines Scandinavian Airlines System and Norwegian Air Shuttle, and for regional aircraft from Western Norway.

Heliports are common at hospitals and oil platforms. The Norwegian Air Ambulance service operates twelve helicopters and nine airplanes.

Regional Aviation

The regional airport service was introduced in the 1960s, with 30 airports being served by short take-off and landing aircraft. These are located mainly in Sogn og Fjordane and Northern Norway, in areas with long distances to large cities and with too little traffic to support commercial flights. The airports, which typically have a 800 metres (2,600 ft) runway, are run by Avinor, while the airplanes are operated based on subsidized public service obligation contracts with the Norwegian Ministry of Transport and Communications. The by far largest contractor is Wideroe with their fleet of de Havilland Canada Dash 8 aircraft, but also Danish Air Transport, Lufttransport and Kato Air have won bids. The flights operate from one or more regional airports to larger hubs; in Oslo,

Bergen, Trondheim, Bodo, Tromso and Kirkenes. One service, to Varoy Airport, is served by helicopter. 1,214,508 passengers passed through the regional airports in 2007.

Rail Transport

The main railway network consists of 4,114 kilometres (2,556 mi) of standard gauge lines, of which 242 kilometres (150 mi) is double track and 64 kilometres (40 mi) high-speed rail (210 km/h) while 62% is electrified at 15 kV 16T! Hz AC. The railways transported 56,827,000 passengers 2,956 million passenger kilometers and 24,783,000 tonnes of cargo 3,414 million tonne kilometers.

The main long-haul network consists of lines from Oslo and westwards along the South Coast to Stavanger and over the mountains to Bergen; and north to Andalsnes and via Trondheim to Bodo. Four lines connect to Sweden, allowing access to the European network. The only high-speed line is Gardermobanen, connecting Oslo to Oslo Airport, Gardermoen, but plans exist to built more high-speed lines in Eastern Norway, and possibly to other parts of Norway. The entire network is owned by the Norwegian National Rail Administration, while all domestic passenger trains except the Airport Express Train are operated by Norges Statsbaner (NSB). Several companies operate freight trains.

Investment in new infrastructure and maintenance is financed through the state budget, and subsidies are provided for passenger train operations. NSB operates long-haul trains, including night trains, regional services and four commuter train systems, around Oslo, Trondheim, Bergen and Stavanger.

Rail Transit

Tramways operate in Oslo and Trondheim, with a system in Bergen under construction. The only rapid transit system is the Oslo T-bane, while the only funicular is in Bergen. The rail transits are operated by the counties, and integrated into the bus transport. In Oslo the two systems make the backbone of the public transport system, giving Oslo by far the highest public transport share of 20%. In 2007, 101 million passengers were transported 490 million passenger kilometers by rail transit.

Road Transport

Norway has a road network of 92,946 kilometres (57,754 mi), of which 72,033 kilometres (44,759 mi) are paved and 664 kilometres (413 mi) are motorway. There are four tiers of road routes; national, county, municipal and private, with only the national roads numbered en route. The most important national routes are part of the European route scheme, and the two most prominent are the E6 going north-south through the entire country, while E39 follows the West Coast. National and county roads are managed by the Norwegian Public Roads Administration.

Motorways exist around the largest cities; many of the larger cities have introduces toll schemes to help finance roads. In 2008, 130 ferry routes remained in service, operated by private companies on contract with the Public Roads Administration. Since the 1970s the heaviest rural investments have been mainland connections to replace the many car ferries that are needed to cross fjords and connect to islands. There are not enough funding through tax money, so these tunnels and bridges are normally financed mainly through toll fees. Some mountain passes have severe snowstorm problems in the winter, so often they have to be closed, or cars have to drive after a snowplough in a column.. The most exposed mountain passes are closed the entire winter.

In 2007 there were 2.6 million automobiles in Norway, or 444 per 1000 residents, an increase of 27% the last ten years—average age was 10.2 years. Road accidents killed 242 people and road transport caused 20% of greenhouse gas emissions. Trucks transported 264 million tonnes 15 billion tonne kilometers.

Bus Transport

Each county is responsible for the public bus and boat transport in their area, (railways, regional airlines and the Coastal Express boat, are financed by the state). Buses transported 290 million passengers 3.7 billion passenger kilometers in 2007. 6,194 buses were in operation during 2007; tickets sales was NOK 3,721 million while bus transport received government subsidies of NOK 3,393 million. Bus, bus and water bus services are normally operated by private companies on contract with the county or their public transport authority (such as Ruter or Vestviken Kollektivtrafikk). Long-haul coach services are operated by various companies, most of whom cooperate through NOR-WAY Bussekspress.

Water Transport

The coastal infrastructure is operated by the Norwegian Coastal Administration, while ports are operated by the municipalities. Norway has 90,000 kilometres (56,000 mi) of shoreline, 400,000 leisure craft and a 715 ships in the merchant marine.

Merchant Marine

Norway is the fifth largest beneficial shipowning country, with 5% of the world's fleet; though a high portion of these are registered in flags of convenience, Norway has 15 million gross tonnes of ships under its flag. The government has created an internal register, the Norwegian International Ship Register (NIS), as a subset of the Norwegian Ship Register; ships on the NIS enjoy many benefits of flags of convenience and do not have to be crewed by Norwegians.

Ferries

Fast ferries operate many places where fjords and islands make it quicker to follow the waterways than the roads; some small islands are served by water buses. Public transport by ship transported eight million passengers 273 million passenger kilometers in 2007. The Coastal Express (known as Hurtigruten) operates daily cruiseferries from Bergen to Kirkenes, calling at 35 ports. International cruiseferries operate from Southern Norway to the United Kingdom, Denmark, Germany and Sweden.

Pipelines

The petroleum and natural gas production on the Norwegian continental shelf uses pipelines to transport produce to processing plants on mainland Norway and other European countries; total length is 9,481 kilometres (5,891 mi). The government-owned Gassco operates all natural gas piplines; in 2006, 88 billion cubic meters were transported, or 15% of European consumption. Due to the low population density, narrow shape and long coastlines, public transport in Norway is less developed than in many European countries, especially outside the cities. As such, Norway has old water transport traditions, but the Norwegian Ministry of Transport and Communications has in recent years implemented rail, road and air transport through numerous subsidiaries in order to develop the country's infrastructure.

Norway's main railway network consists of 4,114 kilometres (2,556 mi) of standard gauge lines, of which 242 kilometres (150 mi) is double track and 64 kilometres (40 mi) high-speed rail (210 km/h) while 62% is electrified at 15 kV 16T! Hz AC. The railways transported 56,827,000 passengers 2,956 million passenger kilometres and 24,783,000 tonnes of cargo 3,414 million tonne kilometres. The entire network is owned by the Norwegian National Rail Administration, while all domestic passenger trains except the Airport Express Train are operated by Norges Statsbaner (NSB). Several companies operate freight trains.

Investment in new infrastructure and maintenance is financed through the state budget, and subsidies are provided for passenger train operations. NSB operates long-haul trains, including night trains, regional services and four commuter train systems, around Oslo, Trondheim, Bergen and Stavanger.

There are approximately 92,946 kilometres (57,754 mi) of road network in Norway, of which 72,033 kilometres (44,759 mi) are paved and 664 kilometres (413 mi) are motorway. There are four tiers of road routes; national, county, municipal and private, with only the national roads numbered en route. The most important national routes are part of the European route scheme, and the two most prominent are the E6 going north-south through the entire country, while E39 follows the West Coast. National and county roads are managed by the Norwegian Public Roads Administration.

Of the 97 airports in Norway, 52 are public, and 46 are operated by the state-owned Avinor. Seven airports have more than one million passengers annually. 41,089,675 passengers passed through Norwegian airports in 2007, of which 13,397,458 were international.

The central gateway by air to Norway is Oslo Airport, Gardermoen, located about 50 kilometres (31 mi) north of Oslo with departures to most European countries and some intercontinental destinations. It is hub for the two major Norwegian airlines Scandinavian Airlines System and Norwegian Air Shuttle, and for regional aircraft from Western Norway.

Education

Higher education in Norway is offered by a range of seven universities, five specialized colleges, 25 university colleges as well as a range of private colleges. Education follows the Bologna process involving Bachelor (3 years), Master (2 years) and PhD (3 years) degrees. Acceptance is offered after finishing upper secondary school with general study competence.

Public education is virtually free, regardless of nationality, with an academic year with two semesters, from August to December and from January to June. The ultimate responsibility for the education lies with the Norwegian Ministry of Education and Research.

Demography

Historical Populations

Year	*Population*	*Percentage*
1500	140,000	—
1665	440,000	214.3%
1735	616,109	40.0%
1801	883,603	43.4%
1855	1,490,047	68.6%
1900	2,240,032	50.3%
1950	3,278,546	46.4%
2000	4,478,497	36.6%
2050?	6,627,000	48.0%

Source: *Statistics Norway.*

Norway's population numbers roughly 4.8 million. Most Norwegians are ethnic Norwegians, a North Germanic people. The Sami people traditionally inhabit central and northern parts of Norway and Sweden, as well as in northern Finland and in Russia on the Kola Peninsula. Another national minority are the Kven people who are the descendants of Finnish speaking people that moved to northern Norway in the 18th up to the 20th century. Both the Sami and the Kven were subjected to a strong assimilation policy by the Norwegian government from the 19th century up to the 1970s.

Because of this "Norwegianization process", many families of Sami or Kven ancestry now self-identify as ethnic Norwegian. This, combined with a long history of co-habitation of the Sami and North Germanic peoples on the Scandinavian peninsula, makes claims about ethnic population statistics less straightforward than is often suggested—particularly in central and northern Norway. Other groups recognized as national minorities of Norway are Jews, Forest Finns, Roma/Gypsies and Romani people/Travellers.

In recent years, immigration has accounted for more than half of Norway's population growth. According to Statistics Norway (SSB), a record 61,200 immigrants arrived in the country in 2007—35% higher than 2006. At the beginning of 2010, there were 552,313 persons in Norway with an immigrant background (i.e. immigrants, or born of immigrant parents), comprising 11.4% of the total population. 210,725 were from Western countries (EU/EFTA, US, Canada, Australia, New Zealand) and 341,588 were from other countries. The largest immigrant groups by country of origin, in order of size, are Poles, Swedes, Pakistanis, Iraqis, Somalis, Germans, Vietnamese, and Danes.

Norwegians of Pakistani descent are the largest visible minority group in Norway, and most of their 31,000 members live around Oslo. The Iraqi immigrant population has shown a large increase in recent years. After the enlargement of the EU in 2004, there has also been an influx of immigrants from Central and Eastern Europe, particularly Poland. The largest increase in 2007 was of immigrants from Poland, Germany, Sweden, Lithuania and Russia.

There are almost 4.7 million Norwegian Americans according to the 2006 U.S. census. The number of Americans of Norwegian descent living in the U.S. today is roughly equal to the current population of Norway. In the 2006 Canadian census, 432,515 Canadian citizens claimed Norwegian ancestry.

Religion

Norwegians were members of the state Church of Norway as of January 1, 2009, a 1% drop compared to the year before and down 2% from two years earlier. Norwegians are registered at baptism as members of the Church of Norway, many remain in the state church to be able to use services such as baptism, confirmation, marriage and burial, rites which have strong cultural standing in Norway.

However, only 20% of Norwegians say that religion occupies an important place in their life (according to a recent Gallup poll), making Norway one of the most secular countries of the world (only in Estonia, Sweden and Denmark were the percentage of people who considered religion to be important lower). In the early 1990s, it was estimated that between 4.7% – 5.3% of Norwegians attended church on a weekly basis. Up to 40% of the membership attends church or religious meetings at least once annually.

According to the most recent Eurobarometer Poll 2005, 32% of Norwegian citizens responded that "they believe there is a god," whereas 47% answered that "they believe there is some sort of spirit or life force" and 17% that "they do not believe there is any sort of spirit, god, or life force." According to Gustafsson and Pettersson (2002), 72% of Norwegians do not believe in a 'personal God.'

Just above 10% of the population is unaffiliated as per 1 January 2009 (as 80.7% were members of the Church of Norway and another 9% or 431 000, were members of religious and life stance communities outside the Church of Norway). Other Christian denominations total about 4.9% of the population, including the Evangelical Lutheran Free Church, the Roman Catholic Church, the Baptists, Pentecostal congregations, the Methodist Church, and Adventists, and others.

Among non-Christian religions, Islam is the largest, representing about 1.5% of the population. It is practiced mainly by Somali, Arab, Albanian, and Turkish immigrants, as well as Norwegians of Pakistani descent. Other religions comprise less than 1% each, including Judaism as well as the Church of Jesus Christ of Latter-day Saints, and Jehovah's Witnesses. Indian immigrants introduced Hinduism to Norway but account for fewer than 5,000 people, or 1% of non-Lutheran Norwegians. There are eleven Buddhist organizations, grouped under the Buddhistforbundet organization, which make up 0.42% of the population. Around 1.5% of Norwegians adhere to the secular

Norwegian Humanist Association. Like other Scandinavian countries, the Norse followed a form of native Germanic paganism known as Norse paganism. By the end of the eleventh century, when Norway had been Christianized, the indigenous Norse religion and practices were prohibited. Remnants of the native religion and beliefs of Norway survive today in the form of names, referential names of cities and locations, the days of the week, and other parts of the everyday language. Modern interest in the old ways has led to a revival of the pagan religious practices in the form of Asatru. The Norwegian Asatrufellesskapet Bifrost formed in 1996; as of 2005, the fellowship has some 200 members. Foreningen Forn Sed formed in 1999 and has been recognized by the Norwegian government as a religious organization.

Parts of the Sami minority retained their shamanistic religion well into the 18th century when they were converted to Christianity by Dano-Norwegian missionaries.

Orthodoxy is the fastest-growing religion in Norway with a rate from 2000 to 2009 of 231.1% compared to Islam's 64.3%.

Language

The North Germanic Norwegian language has two official written forms, *Bokmal* and *Nynorsk*. Both of them are recognized as official languages, in that they are both used in public administration, in schools, churches, and media, but Bokmal is used by the vast majority, about 85–90%. Around 95% of the population speak Norwegian as their native tongue, although many speak dialects that may differ significantly from the written language. In general, most Norwegian dialects are inter-intelligible, although some may require significant efforts on the part of a listener to understand. Several Finno-Ugric Sami languages are spoken and written throughout the country, especially in the north, by the *Sami people*. Speakers have a right to get education in Sami language no matter where they are living and receive communication from the government in various Sami languages. The Kven minority speak the Finno-Ugric Kven language/Finnish. There is advocacy for making Norwegian Sign Language an official Norwegian language.

In the 19th and 20th century, Norwegian language was subject to strong political and cultural controversy, which led to the creation of Nynorsk in the 19th century and to the formation of alternative spelling standards in the 20th century, notably the Riksmal standard, which is more conservative (that is, more similar to Danish) than Bokmal.

Norwegian is similar to the other languages in Scandinavia, Swedish and Danish. All three languages are mutually intelligible and can be, and commonly are, employed in communication between inhabitants of the Scandinavian countries. As a result of the cooperation within the Nordic Council, inhabitants of all Nordic countries, including Iceland and Finland, have the right to communicate with the Norwegian authorities in their own language.

Any Norwegian student who is a child of immigrant parents is encouraged to learn the Norwegian language. The Norwegian government offers language instructional courses for immigrants wishing to obtain Norwegian citizenship. However, immigrants are not required to learn Norwegian in order to gain Norwegian citizenship, nor are their children.

The main foreign language taught in Norwegian elementary school is English. The majority of the population is fluent in English, especially those born after World War II. German, French and Spanish are also commonly taught as a second or, more often, third language. Russian, Japanese, Italian, Latin and rarely Chinese (Mandarin) are available in some schools, mostly in the cities. Traditionally, English, German and French were considered the main foreign languages in Norway. These languages, for instance, had been used on Norwegian passports until the 1990s, and university students have a general right to use these languages when submitting their theses.

Culture

Norwegian culture is closely linked to the country's history and geography. The unique Norwegian farm culture, sustained to this day, has resulted not only from scarce resources and a harsh climate but also from ancient property laws. In the 18th century, it brought about a strong romantic nationalistic movement, which is still visible in the Norwegian language and media. In the 19th century, Norwegian culture blossomed as efforts continued to achieve an independent identity in the areas of literature, art and music. This continues today in the performing arts and as a result of government support for exhibitions, cultural projects and artwork.

Norway is an early adopter of women's rights, minority rights, and LGBT rights. For example, in 1990 Norway was the first country to recognize the ILO-convention 169 about indigenous people, in 1993 Norway became the second country to legalize civil union partnerships for same-sex couples, and on January 1, 2009, Norway became the sixth country to grant full marriage equality to same-sex couples.

Cuisine

Norway's culinary traditions show the influence of long seafaring and farming traditions with salmon (fresh and cured), herring (pickled or marinated), trout, codfish and other seafood balanced by cheeses, dairy products and breads (predominantly dark/darker). Lefse is a common Norwegian potato flatbread, common around Christmas.

Performing Arts

Film

Not until fairly recently has the Norwegian cinema received international recognition, but as early as 1951 a documentary film of the *Kon-Tiki* expedition won an Oscar Academy Award. In 1959, Arne Skouen's *Nine Lives* was nominated, but failed to win. Another notable film is *Flaklypa Grand Prix* (English: *Pinchcliffe Grand Prix*), an animated feature film directed by Ivo Caprino. The film was released in 1975 and is based on characters from Norwegian cartoonist Kjell Aukrust. It is the most widely seen Norwegian film of all time.

There was however a real breakthrough in 1987 with Nils Gaup's *Pathfinder* which told the story of the Sami. It was nominated for an Oscar and was a huge international success. Berit Nesheim's *The Other Side of Sunday* was also nominated for an Oscar in 1997.

Since the 1990s, the film industry has thrived with up to 20 feature films each year. Particular successes were *Kristin Lavransdatter*, *The Telegraphist* and *Gurin with the Foxtail*. Knut Erik Jensen was among the more successful new directors together with Erik Skjoldbjaerg remembered for *Insomnia*.

In late 2008, the movie *Max Manus* opened at Norwegian theatres. The movie was a WW2 drama, telling the story of the Norwegian resistance hero Max Manus who led many successful sabotage operations against the German occupation. The movie became the highest grossing Norwegian movie ever.

Music

Along with the classical music of romantic composer Edvard Grieg and the modern music of Arne Nordheim, Norwegian black metal has become something of an export article in recent years.

Norway's classical performers include Leif Ove Andsnes, one of the world's more famous pianists, and Truls Mork, an outstanding cellist.

The jazz scene in Norway is also thriving. Jan Garbarek, Terje Rypdal, Mari Boine, Arild Andersen, and Bugge Wesseltoft are internationally recognized while Paal Nilssen-Love, Supersilent, Jaga Jazzist and Wibutee are becoming world-class artists of the younger generation.

Norway has a strong folk music tradition which remains popular to this day. Among the most prominent folk musicians are Hardanger fiddlers Andrea Een, Olav Jorgen Hegge and Annbjorg Lien, vocalists Agnes Buen Garnas, Kirsten Braten Berg and Odd Nordstoga.

Since the 1990s, Norway's biggest cultural export is black metal. The lo-fi, dark and raw form of heavy metal exploded in Norway during the 90s and launched the worldwide acclaimed careers of bands such as Gorgoroth, Mayhem, Burzum, Emperor, Darkthrone and Immortal, as well as later bands such as Dimmu Borgir. This development has been somewhat well-received for the musical value, but many events that took place in the early 1990s as a result of the black metal movement caused quite a panic amongst the Norwegian citizens at large.

Literature

History of Norwegian literature starts with the pagan Eddaic poems and skaldic verse of the 9th and 10th centuries with poets such as Bragi Boddason and Eyvindr Skaldaspillir. The arrival of Christianity around the year 1000 brought Norway into contact with European mediaeval learning, hagiography and history writing. Merged with native oral tradition and Icelandic influence this was to flower into an active period of literature production in the late 12th and early 13th centuries. Major works of that period include *Historia Norwegie*, *Thidreks saga* and *Konungs skuggsja*.

Little Norwegian literature came out of the period of the Scandinavian Union and the subsequent Dano-Norwegian union (1387—1814), with some notable exceptions such as Petter Dass and Ludvig Holberg. In his play Peer Gynt, Ibsen characterized this period as "Twice two hundred years of darkness/brooded o'er the race of monkeys", although the latter line is not as frequently quoted as the former. During the union with Denmark, written Norwegian was replaced by Danish.

Two major events precipitated a major resurgence in Norwegian literature. In 1811 a Norwegian university was established in Christiania. Seized by the spirit of revolution following the American and French Revolutions, the Norwegians signed their first Constitution in 1814. Soon, the cultural backwater that was Norway brought forth a series of strong authors recognized first in Scandinavia, and then worldwide; among them were Henrik Wergeland, Peter Asbjornsen, Jorgen Moe and Camilla Collett.

By the late 19th century, in the Golden Age of Norwegian literature, the so-called *Great Four* emerged: Henrik Ibsen, Bjornstjerne Bjornson, Alexander Kielland, and Jonas Lie. Bjornson's "peasant novels", such as "En glad gutt" (A Happy Boy) and "Synnøve Solbakken" are typical of the national romanticism of their day, whereas Kielland's novels and short stories are mostly realistic. Although an important contributor to early Norwegian romantic nationalism (especially the ironic Peer Gynt), Henrik Ibsen's fame rests primarily on his pioneering realistic dramas such *The Wild Duck* and *A Doll's House*, many of which caused moral uproar because of their candid portrayals of the middle classes.

In the 20th century, three Norwegian novelists were awarded the Nobel prize in literature: Bjornstjerne Bjornson in 1903, Knut Hamsun for the book "Markens grade" ("Growth of the Soil") in 1920, and Sigrid Undset in 1928. Further important contributions to Norwegian literature were made by writers like Dag Solstad, Jon Fosse, Cora Sandel, Olav Duun,

Olav H. Hauge, Gunvor Hofmo, Stein Mehren, Kjell Askildsen, Hans Herbjornsrud, Aksel Sandemose, Bergljot Hobæk Haff, Jostein Gaarder, Erik Fosnes Hansen, Jens Bjorneboe, Kjartan Flogstad, Lars Saabye Christensen, Johan Borgen, Herbjørg Wassmo, Jan Erik Vold, Rolf Jacobsen, Olaf Bull, Jan Kjærstad, Georg Johannesen, Tarjei Vesaas, Sigurd Hoel, Arnulf Overland and Johan Falkberget.

Architecture

Norway has always had a tradition of building in wood. Indeed, many of today's most interesting new buildings are made of wood, reflecting the strong appeal that this material continues to hold for Norwegian designers and builders.

Norway's conversion to Christianity some 1,000 years ago led to the introduction of stonework architecture, beginning with the construction of Nidaros Cathedral in Trondheim.

In the early Middle Ages, stave churches were constructed throughout Norway. Many of them remain to this day and represent Norway's most important contribution to architectural history. A fine example is The Stave Church at Urnes which is now on UNESCO's World Heritage List. Another notable example of wooden architecture is the Bryggen Wharf in Bergen, consisting of a row of narrow wooden structures along the quayside. In the 17th century, under the Danish monarchy, cities such as Kongsberg with its Baroque church and Roros with its wooden buildings were established.

After Norway's union with Denmark was dissolved in 1814, Oslo became the capital. Architect Christian H. Grosch designed the oldest parts of the University of Oslo, the Oslo Stock Exchange, and many other buildings and churches.

At the beginning of the 20th century, the city of Alesund was rebuilt in the Art Nouveau style. The 1930s, when functionalism dominated, became a strong period for Norwegian architecture, but it is only in recent decades that Norwegian architects have truly achieved international renown. One of the most striking modern buildings in Norway is the Sami Parliament in Karasjohka designed by Stein Halvorson and Christian Sundby. Its debating chamber is an abstract timber version of a Lavvo, the traditional tent used by the nomadic Sami people.

Art

For an extended period, the Norwegian art scene was dominated by artwork from Germany and Holland as well as by the influence of Copenhagen. It was in the 19th century that a truly Norwegian era began, first with portraits, later with even more impressive landscapes. Johan Christian Dahl (1788–1857), originally from the Dresden school, eventually returned to paint the landscapes of western Norway, defining Norwegian painting for the first time."

Norway's new-found independence from Denmark encouraged painters to develop their Norwegian identity, especially with landscape painting by artists such as Kitty Kielland, a female painter who studied under Gude; Harriet Backer, 1845–1932, another pioneer among female artists, influenced by impressionism. Frits Thaulow, an impressionist, was influenced by the art scene in Paris as was Christian Krohg, a realist painter, famous for his paintings of prostitutes.

Of particular note is Edvard Munch, a symbolist/ expressionist painter who became world famous for The Scream which is said to represent the anxiety of modern man.

Other artists of note include Harald Sohlberg, a neo-romantic painter remembered for his paintings of Roros; and Odd Nerdrum, a figurative painter who maintains his work is not art but kitsch.

9

Scandinavia and Finland

Scandinavia is a region in northern Europe that includes Denmark and the Scandinavian Peninsula's two nations, Norway and Sweden. Finland is sometimes considered a Scandinavian country in common English usage, and Iceland and the Faroe Islands are sometimes also included, although this usage is considered incorrect in Scandinavia. The term Nordic countries refers to Denmark, Norway and Sweden as well as Finland and Iceland, and associated territories.

Scandinavia extends to the north of the Arctic Circle, but has relatively mild weather for its latitude, owing to the Gulf Stream. Much of the Scandinavian mountains have an alpine tundra climate. There are many lakes and moraines, legacies of an ice age about 10,000 years ago. The northern regions of Scandinavia are home to the indigenous Sami people.

Scandinavia has, despite many wars over the years since the formation of modern nation-states in the 10th century, been politically and culturally close. The constellations and alliances, however, have shifted over the centuries. For all of the 15th century, Scandinavia was united in the Kalmar Union. In the 19th century a new political union was proposed, but it did not take place when Denmark was denied key military support in a conflict with Prussia. Today, the nations cooperate mainly in the European Union or the Nordic Council.

The Danish, Faroese, Icelandic, Norwegian and Swedish languages are linguistically classified as North Germanic languages (also called Scandinavian languages), while the Finnish and Sami languages are classified as members of the

Finno-Lappic group of the Uralic language family, unrelated to the Scandinavian languages. Danish, Norwegian and Swedish form a mutually intelligible dialect continuum, which is seen as a defining characteristic of Scandinavia as a modern cultural and linguistic entity.

Terminology and Usage

In English, casual and unofficial usage is to include the countries of Denmark, Norway, Sweden and often Finland and Iceland in the term "Scandinavia", but the term is usually used to only mean Denmark, Norway, and Sweden within those countries creating some ambiguity in even English reference sources.

The larger region that some English-speaking nations refer to as "Scandinavia" is known by the countries concerned as Norden, or the Nordic countries, a political entity as well as cultural region where the ties between the countries are not merely historical and cultural, but based on official membership in the Nordic Council. Some American-English dictionaries, such as Merriam-Webster Online Dictionary, do not include the names "Nordic countries" or "Nordic Council". Merriam-Webster Online Dictionary instead defines Nordic as an adjective dated to 1898 with the meaning "of or relating to the Germanic peoples of northern Europe and especially of Scandinavia."

The use of the name Scandinavia as a convenient general term for the three kingdoms of Denmark, Norway, and Sweden is also fairly recent; according to some historians, it was adopted and introduced in the 18th century, at a time when the ideas about a common heritage started to appear and develop into early literary and linguistic Scandinavism. Before this time, the term *Scandinavia* was familiar mainly to classical scholars through Pliny the Elder's writings, and was used vaguely for Scania and the southern region of the peninsula.

As a political term, "Scandinavia" was first used by students agitating for Pan-Scandinavianism in the 1830s. The popular usage of the term in Sweden, Denmark and Norway as a unifying concept became established in the 19th century through poems such Hans Christian Andersen's "I am a Scandinavian" of 1839. After a visit to Sweden, Andersen became a supporter of early political Scandinavism and in a letter describing the poem to a friend, he wrote: "All at once I understood how related the Swedes, the Danes and the Norwegians are, and with this feeling I wrote the poem immediately after my return: 'We are one people, we are called Scandinavians!'". The historic popular usage is also reflected in the name chosen for the shared, multi-national airline, Scandinavian Airlines System, a carrier originally owned jointly by the governments of the three countries, along with private investors.

Finland

The clearest example of the use of the term "Scandinavia" as a political and cultural construct is the unique position of Finland, based largely on it having been part of Sweden, thus to much of the world properly associating Finland with all of Scandinavia. But the creation of a Finnish identity is unique in the region in that it was formed in relation to two different imperial models, the Swedish and the Russian, as described by the University of Jyvaskyla based editorial board of the Finnish journal *Yearbook of Political Thought and Conceptual History*;

The construction of a specific Finnish polity is the result of successful decolonization. The politico-cultural location of Finland is a moving one. It has shifted from being a province in the Swedish Empire to an autonomous unit in 'Eastern' Europe, then to an independent state in 'Northern' Europe or 'Scandinavia'. After joining the European Union, Finland has recently been included in 'Western Europe'.

The term is, therefore, often defined according to the conventions of the cultures that lay claim to the term in their own usage.

Cultural and Tourism Promotional Organizations

Various promotional agencies of the Nordic countries in the United States (such as The American-Scandinavian Foundation, established in 1910 by the Danish-American industrialist Niels Poulsen) serve to promote market and tourism interests in the region. Today, the five Nordic Heads of State act as the organization's patrons and according to the official statement by the organization, its mission is "to promote the Nordic region as a whole while increasing the visibility of Denmark, Finland, Iceland, Norway and Sweden in New York City and the United States." The official tourist boards of Scandinavia sometimes cooperate under one umbrella, such as the Scandinavian Tourist Board. The cooperation was introduced for the Asian market in 1986, when the Swedish national tourist board joined the Danish national tourist board to coordinate international promotions of the two countries. Norway entered one year later. All five Nordic countries participate in the joint promotional efforts in the United States through the Scandinavian Tourist Board of North America.

Use of Nordic Countries *vs.* Scandinavia

While the term *Scandinavia* is commonly used for Denmark, Norway and Sweden, the term *the Nordic countries* is used unambiguously for Denmark, Norway, Sweden, Finland, and Iceland, including their associated territories (Greenland, the Faroe Islands, and the Aland Islands). Scandinavia can thus be considered a subset of the Nordic countries.

In addition to the mainland countries of:

- Denmark (Constitutional monarchy with a Parliamentary system — an autonomous constituent country within the Danish Realm)
- Norway (Constitutional monarchy with a Parliamentary system, independent since 1905, also including Svalbard)
- Sweden (Constitutional monarchy with a Parliamentary system, independent kingdom since 1523).

The Nordic countries consist of;

- Finland (Semi-presidential republic, independent since 1917)
- Iceland (Parliamentary republic, independent since 1918, although in a union with and a dependency of Denmark until 1944).
- Faroe Islands (an autonomous constituent country within the Danish realm, self-governed since 1948)
- Greenland (an autonomous constituent country within the Danish realm, self-governed since 1979)
- Aland Islands (an autonomous province of Finland since 1920).

Estonia has applied for membership in the Nordic Council, referring to its cultural heritage and close linguistic links to

Finland, although normally Estonia is regarded as one of the Baltic countries. All Baltic states have shared historical events with the Nordic countries, including Scandinavia, during the centuries.

Etymology

Scandinavia and Scania (*Skane*) are considered to have the same etymology. Both terms are thought to be derived from the Germanic root **Skaoin-awjo*, which appears later in Old English as *Scedenig* and in Old Norse as *Skaney*. The earliest identified source for the name Scandinavia is Pliny the Elder's Natural History, dated to the 1st century AD.

Various references to the region can also be found in Pytheas, Pomponius Mela, Tacitus, Ptolemy, Procopius and Jordanes. It is believed that the name used by Pliny may be of West Germanic origin, originally denoting Scania. According to some scholars, the Germanic stem can be reconstructed as **Skaoan*-meaning "danger" or "damage" (English *scathing*, German *Schaden*). The second segment of the name has been reconstructed as **awjo*, meaning "land on the water" or "island". The name Scandinavia would then mean "dangerous island", which is considered to be a reference to the treacherous sandbanks surrounding Scania. Skanor in Scania, with its long Falsterbo reef, has the same stem (*skan*) combined with-*ör*, which means "sandbanks".

In the reconstructed Germanic root **Skaðin-awjo* (the edh represented in Latin by t or d), the first segment is sometimes considered more uncertain than the second segment. The American Heritage Dictionary derives the second segment from Proto-Indo-European **akwa-*, "water", in the sense of "watery land". Gothic *saiws*, "lake" is one of the Germanic groups which include English *sea* and German *See*. However, according to the Indo-European Dictionary (IEED), a research project of the Department of Comparative Indo-European Linguistics at Leiden University, the second segment may not have an Indo-European etymology. The IEED states that Uralic evidence has long been recognized for this segment, namely the Finnish *saivo* ("'transparent place in the sea'") and the Northern Sami *saivva* ("'(holy) lake, idol'"). Some scholars have found a parallel between the Uralic evidence and the area's old mythology and belief systems, where the soul of mankind is believed to dwell in water until birth and return there after death. IEED lists a Germanic reconstruction that indicates a similar connection to metaphysics, namely **saiwa-lo* ("soul"), appearing as *saiwala* in Gothic and *scle* in Old Frisian.

Pliny the Elder's Descriptions

Pliny's descriptions of *Scatinavia* and surrounding areas are not always easy to decipher, even though his writing of geography was what he considered a "*clarior fama*" ("a clearer story"). Writing in the capacity of a Roman admiral, he introduces the northern region by declaring to his Roman readers that there are 23 islands "*Romanis armis cognitae*" ("known to Roman arms") in this area. According to Pliny, the "*clarissima*" ("most famous") of the region's islands is *Scatinavia*, of unknown size. There live the Hilleviones. The belief that Scandinavia was an island became widespread among classical authors during the first century and dominated descriptions of Scandinavia in classical texts during the centuries that followed.

Pliny begins his description of the route to *Scatinavia* by referring to the mountain of Saevo (*mons Saevo ibi*), the Codanus Bay (*Codanus sinus*) and the Cimbrian promontory. The geographical features have been identified in various ways; by some scholars "Saevo" is thought to be the mountainous Norwegian coast at the entrance to Skagerrak and the Cimbrian peninsula is thought to be Skagen, the north tip of Jutland, Denmark. As described, Saevo and Scatinavia can also be the same place.

Pliny mentions Scandinavia one more time: in Book VIII he says that the animal called *achlis* (given in the accusative, *achlin*, which is not Latin), was born on the island of Scandinavia. The animal grazes, has a big upper lip and some mythical attributes.

The name "Scandia", later used as a synonym for Scandinavia, also appears in Pliny's Naturalis Historia, but is used for a group of Northern European islands which he locates north of Britannia. "Scandia" thus does not appear to be denoting the island Scadinavia in Pliny's text. The idea that "*Scadinavia*" may have been one of the "*Scandiae*" islands was instead introduced by Ptolemy (c.90 – c.168 AD), a mathematician, geographer and astrologer of Roman Egypt. He used the name "*Skandia*" for the biggest, most easterly of the three "*Scandiai*" islands, which according to him were all located east of Jutland.

Neither Pliny's nor Ptolemy's lists of Scandinavian tribes include the Suiones mentioned by Tacitus. Some early Swedish scholars of the Swedish Hyperborean school and of the 19th-century romantic nationalism period proceeded to synthesize the different versions by inserting references to the Suiones, arguing that they must have been referred to in the original texts and obscured over time by spelling mistakes or various alterations.

Germanic Reconstruction

The Latin names in Pliny's text gave rise to different forms in medieval Germanic texts. In Jordanes' history of the Goths (AD 551) the form *Scandza* is used for their original home, separated by sea from the land of Europe. Where Jordanes meant to locate this quasi-legendary island is still a hotly debated issue, both in scholarly discussions and in the nationalistic discourse of various European countries. The form *Scadinavia* as the original home of the Langobards appears in Paulus Diaconus' *Historia Langobardorum*; in other versions of *Historia Langobardorum* appear the forms *Scadan*, *Scandanan*, *Scadanan* and *Scatenauge*. Frankish sources used *Sconaowe* and Aethelweard, an Anglo-Saxon historian, used *Scani*. In Beowulf, the forms *Scedenige* and *Scedeland* are used, while the Alfredian translation of Orosius and Wulfstan's travel accounts used the Old English *Sconeg*.

The first segment in "Scandinavia" is also sometimes attributed to Norse mythology, namely the Scandinavian giantess Skaoi (Skade).

Sami Etymology

The earliest Sami yoik texts written down refer to the world as *Skadesi-suolo* (north-Sami) and *Skaosual* (east-Sami), meaning "Skaoi's island" (Svennung 1963). Svennung considers the Sami name to have been introduced as a loan word from the North Germanic languages; "Skaði" is the giant stepmother of Freyr and Freyja in Norse mythology. It has been suggested that Skaði to some extent is modeled on a Sami woman. The name for Skade's father Thjazi is known in Sami as *Eahci*, "the waterman", and her son with Odin, Saeming, can be interpreted as a descendent of *Saam* the Sami population (Mundel 2000), (Steinsland 1991). Older joik texts give evidence of the old Sami belief about living on an island and state that the wolf is known as *suolu gievra*, meaning "the strong one on the island". The Sami place name *Sulliideielbma* means "the island's threshold" and *Suoloeielgi* means "the island's back".

In recent substrate studies, Sami linguists have examined the initial cluster sk-in words used in Sami and concluded that sk-is a phonotactic structure of non-native origin.

Other Etymologies

Scadin-can be segmented various ways to obtain various Indo-European meanings: scand-or scad-in-, scan-or sca-din, scandin or scadin-. These segmentations have resulted in a number of possible etymologies, such as "climbing island" (*scand-), "island of the Scythian people", "island of the woodland of *sca-".

Another possibility is that all or part of the segments of the name came from the indigenous Mesolithic people inhabiting the region. Today Scandinavia is a peninsula, but between approximately 10,300 BP and 9,500 BP, the southern part of Scandinavia was an island separated from the northern peninsula, with water exiting the Baltic Sea through the area where Stockholm is now located.

Some Basque scholars have presented the idea that the segment *sk* that appears in *Ska"inaujàin is connected to the name for the Euzko peoples, akin to Basques, that populated Paleolithic Europe. According to some of these intellectuals, the Scandinavians share some genetic markers with the Basque people.

The name of the Scandinavian mountain range, *Skanderna* in Swedish, was artificially derived from *Skandinavien* in the 19th century, in analogy with *Alperna* for the Alps. The commonly used names are *bergen* or *fjällen*; both names meaning "the mountains".

Geography

The geography of Scandinavia is extremely varied. Notable are the Norwegian fjords, the Scandinavian Mountains, the flat, low areas in Denmark, and the archipelagos of Sweden and Norway. Sweden, and to an even greater extent Finland, has many lakes and moraines, legacies of the ice age.

The climate varies from north to south and from west to east; a marine west coast climate (Cfb) typical of western Europe dominates in Denmark, southernmost part of Sweden and along the west coast of Norway reaching north to 65°N, with orographic lift giving more ;mm/year precipitation (<5000 mm) in some areas in western Norway.

The central part – from Oslo to Stockholm – has a humid continental climate (Dfb), which gradually gives way to subarctic climate (Dfc) further north and cool marine west coast climate (Cfc) along the northwestern coast. A small area along the northern coast east of the North Cape has tundra climate (Et) as a result of a lack of summer warmth. The Scandinavian Mountains block the mild and moist air coming from the southwest, thus northern Sweden and Finnmarksvidda plateau in Norway receive little precipitation and have cold winters. Large areas in the Scandinavian mountains have alpine tundra climate.

The warmest temperature ever recorded in Scandinavia is 38.0 °C in Malilla (Sweden). The coldest temperature ever recorded is "52.6 °C in Vuoggatjalme (Sweden). The warmest month on record was July 1901 in Oslo, with a mean (24hr) of 22.7 °C, and the coldest month was February 1985 in Vittangi (Sweden) with a mean of-27.2 °C..

Southwesterly winds further warmed by foehn wind can give warm temperatures in narrow Norwegian fjords in winter; Tafjord has recorded 17.9 °C in January and Sunndal 18.9 °C in February.

Geography of Denmark

Denmark is located in Western Europe (it is one of the Nordic countries) on the Jutland peninsula and several islands in the Baltic sea. It sidelines both the (Baltic Sea) and the North Sea along its 7,987 km coastline. Its size is comparable to that of Nova Scotia. Denmark has a 68 km border with Germany. Denmark experiences a temperate climate. This means that the winters are mild and windy and the summers are cool. The local terrain is generally flat with a few gently rolling plains. The territory of Denmark includes the island of Bornholm in the Baltic Sea and the rest of metropolitan Denmark, but excludes the Faroe Islands and Greenland. Its position gives Denmark complete control of the Danish Straits (Skagerrak and Kattegat) linking Baltic and North Seas. The country's natural resources include Petroleum, natural gas, fish, salt, limestone, stone, gravel and sand.

Environment

Land use

- Arable land: 60%
- Permanent crops: 0%
- Permanent pastures: 5%
- Forests and woodland: 10%
- Other: 25% (1993 est.)

Irrigated land: 4 350 km² (1993 est.)

Natural Hazards

- Flooding is a threat in some areas of the country.

Current Issues

- Air pollution, principally from vehicle and power plant emissions
- Nitrogen and phosphorus pollution of the North Sea
- Drinking and surface water becoming polluted from animal wastes and pesticides.

International Agreements

- Party to: Air Pollution, Air Pollution-Nitrogen Oxides,

Air Pollution-Sulphur 85, Air Pollution-Sulphur 94, Air Pollution-Volatile Organic Compounds, Antarctic Treaty, Biodiversity, Climate Change, Desertification, Endangered Species, Environmental Modification, Hazardous Wastes, Law of the Sea, Marine Dumping, Marine Life Conservation, Nuclear Test Ban, Ozone Layer Protection, Ship Pollution, Tropical Timber 83, Tropical Timber 94, Wetlands, Whaling

Signed, but not ratified: Air Pollution-Persistent Organic Pollutants, Antarctic-Environmental Protocol, Climate Change-Kyoto Protocol.

Transnational Issues

Maritime Claims

- Contiguous zone: 24 nmi (44.4 km; 27.6 mi)
- Continental shelf: 200-m depth or to the depth of exploitation
- Exclusive economic zone: 200 nmi (370.4 km; 230.2 mi)
- Territorial sea: 12 nmi (22.2 km; 13.8 mi).

Other Issues

- Iceland disputes the Faroe Islands' fisheries median line.
- Iceland, the United Kingdom, and Ireland dispute Denmark's claim that the Faroe Islands' continental shelf extends beyond 200 nautical miles or about 370 km.
- The Faroe Islands continue to study proposals for full independence.
- Uncontested sovereignty dispute with Canada over Hans Island in the Kennedy Channel between Ellesmere Island and Greenland.
- Denmark and Poland have failed to reach an agreement settling the boundary between the two countries.
- Denmark is currently investigating the extent of the continental shelf of Greenland, in the hope that Greenland's Exclusive Economic Zone can be expanded. One of the areas investigated is the geographical North Pole.

Geography of Finland

The geography of Finland differs from that of other Nordic countries. Bordering the Baltic Sea, Gulf of Bothnia, and Gulf of Finland, as well as Sweden, Norway, and Russia, Finland is the northernmost country on the European continent. Although other countries have points extending north, virtually all of Finland is north of 60 degrees north latitude; nearly a quarter of the land area and fully one-third of the latitudinal extent of the country lie north of the Arctic Circle.

Size, External Boundaries, and Geology

Finland's total area is 337,030 km^2 (130,128 sq mi). Finland is the seventh largest country in Europe after Russia, Ukraine, France, Spain, Sweden and Germany. Of this area 10% is water, 69% forest, 8% cultivated land and 13% other.

The most predominant influences on Finland's geography were the continental glaciers that scoured and gouged the country's surface. When the glaciers receded about 10,000 years ago, they left behind moraines, drumlins, and eskers. Other indications of their presence are the thousands of lakes they helped to form in the southern part of the country. The force of the moving ice sheets gouged the lake beds, and meltwaters helped to fill them. The recession of the glaciers is so recent (in geologic terms) that modern-day drainage patterns are immature and poorly established. The direction of glacial advance and recession set the alignment of the lakes and streams in a general northeast to southwest lineation. The two Salpausselka Ridges, which run parallel to each other about twenty-five kilometers apart, are the terminal moraines. At their greatest height they reach an elevation of about 200 meters, the highest point in southern Finland.

As a whole, the shape of Finland's boundaries resembles a figure of an one-armed human. In Finnish, parallels are drawn between the figure and the national personification of Finland – Finnish Maiden (*Suomi-neito*) – and the country as a whole can be referrerd in the Finnish language by her name. Even in official context the area around Enontekio in northwestern part of the country between Sweden and Norway can be refererred to as the "Arm" (*kasivarsi*). After the Continuation War Finland lost major land areas to Russia in the Moscow Armistice of 1944, and the figure was said to have lost the other of her arms, as well as a hem of her "skirt".

Landform Regions

Many countries of the world can be divided into distinct geographic regions, in each of which some physical characteristic is dominant, almost to the exclusion of others. In Finland, the same physical characteristics are common to each of the four geographic regions into which the country is divided. Regional differences in Finland lie, therefore, in subtle combinations of physical qualities. In archipelago Finland, rock and water are dominant. Coastal Finland consists of broad clay plains where agriculture plays a leading role. The interior Finnish lake district supports extensive forests. Upland Finland is covered by Arctic scrub.

Archipelago Finland, consisting of thousands of islands and skerries, extends from the southwestern coast out into the Baltic Sea. It includes the strategically significant Aland Islands, positioned at the entrance to the Gulf of Bothnia. After World War I, both Finland and Sweden laid claim to the islands, which are culturally more Swedish than Finnish. For strategic reasons, however, the League of Nations awarded the Aland Islands to Finland in 1921. A principal reason for this decision was that, during the winter, the islands are physically linked to Finland by the frozen waters of the sea and are thus essential for the country's defense. These forest-covered and bare bedrock islands were, and continue to be, formed by the process of uplift following the last glaciation.

The rest of the country is also still emerging from the sea. The weight of the continental glaciers depressed the land over which they moved, and even now, a hundred centuries after their recession, Finland is rising up from this great load through the process of isostatic rebound. In the south and the southwest, this process is occurring slowly, at a rate of twenty-five to thirty centimetres a century. Farther north in the Ostrobothnia area, uplift is more rapid, it amounts to eighty or ninety centimetres a century. The process also means that Finland is growing about seven square kilometers yearly as land emerges from the sea.

Coastal Finland consists of broad clay plains extending from the coast inland, for no more than 100 kilometers. These plains slope southward from the morainic Salpausselka Ridges in southern Finland. Along the Gulf of Bothnia coast, the plains slope southwest from upland areas. The land of coastal Finland is used for agriculture and dairy farming.

The interior lake district is the largest geographic region, and it is perhaps what most foreigners think of when they imagine Finland. The district is bounded to the south by the Salpausselka Ridges. Behind the ridges extend networks of thousands of lakes separated by hilly forested countryside. This landscape continues to the east and extends into Russia. As a consequence, there is no natural border between the two countries. Because no set definition of what constitutes a lake and no procedures for counting the number of lakes exist, it has been impossible to ascertain exactly how many lakes the region has. There are, however, at least 55,000 lakes that are 200 or more meters wide. The largest is Lake Saimaa, which, with a surface area of more than 4,400 square kilometers, is the fifth largest lake in Europe. The deepest lake has a depth of only 100 meters; the depth of the average lake is 7 meters. Because they are shallow, these many lakes contain only slightly more water than Finland's annual rainfall. The hilly, forest-covered landscape of the lake plateau is dominated by drumlins and by long sinuous eskers, both glacial remnants.

Upland Finland extends beyond the Arctic Circle. The extreme north of this region is known as Lapland. The highest points in upland Finland reach an elevation of about 1,000 meters, and they are found in the Kilpisjarvi area of the Scandinavian Keel Ridge. In the southern upland region the hills are undulating, while in the north they are rugged. Much of upland Finland is not mountainous, but consists of bogs.

Finland's longest and most impressive rivers are in the north. The Kemijoki has the largest network of tributaries. Farther south the Oulujoki drains the beginning of the north country.

Climate

Latitude is the principal influence on Finland's climate. Because of Finland's northern location, winter is the longest season. On the average, winter lasts from early December to mid March in the archipelago and the southwestern coast and from early October to early May in Lapland. This means that southern portions of the country are snow-covered about three to four months of the year and the northern, about seven months. The long winter causes about half of the annual 500 to 600 millimetres (19.7 to 23.6 in) of precipitation in the north to fall as snow. Precipitation in the south amounts to about 600 to 700 millimetres (23.6 to 27.6 in) annually. Like that of the north, it occurs all through the year, though not so much of it is snow.

The Atlantic Ocean to the west and the Eurasian continent to the east interact to modify the climate of the country. The warm waters of the Gulf Stream and the North Atlantic Drift Current, which warm Norway and Sweden, also warm Finland. Westerly winds bring the warm air currents into the Baltic areas and to the country's shores, moderating winter temperatures, especially in the south. These winds, because of clouds associated with weather systems accompanying the westerlies, also decrease the amount of sunshine received during the summer. By contrast, the continental high pressure system situated over the Eurasian continent counteracts the maritime influences, causing severe winters and occasionally high temperatures in the summer.

The highest ever recorded temperature is 35.9 °C (96.6 °F) (Turku, July 1914). The lowest, "51.5 °C ("61 °F) (Kittila, January 1999). The annual middle temperature is relatively high in the southwestern part of the country (5.0 °C/41.0 °F to 7.5 °C/45.5 °F), with quite mild winters and warm summers, and low in the northeastern part of Lapland (0 °C/32 °F to "4 °C/24.8 °F).

Area and Boundaries

Area:

Total: 337,030 km^2

Land: 305,470 km^2

Water: 31,560 km^2

Area-comparative: slightly smaller than Germany.

Land boundaries:

Total: 2,628 km.

Border countries: Norway 729 km, Sweden 586 km, Russia 1,313 km.

Coastline: 1,126 km (excludes islands and coastal indentations).

Maritime claims:

Contiguous zone: 6 nmi (11.1 km; 6.9 mi).

Continental shelf: 200 m depth or to the depth of exploitation.

Exclusive fishing zone: 12 nmi (22.2 km; 13.8 mi).

Territorial sea: 12 nmi (22.2 km; 13.8 mi), 3 nmi (5.56 km; 3.45 mi) in the Gulf of Finland.

Elevation extremes:

Lowest point: Baltic Sea 0 m.

Highest point: Haltitunturi 1,328 m.

Resources and Land Use

Natural resources: timber, copper, zinc, iron ore, silver

Land use:

Arable land:	8%
Permanent crops:	0%
Permanent pastures:	0%
Forests and woodland:	76%
Other:	16% (1993 est.)

Irrigated land: 640 km^2 (1993 est.)

Environmental Concerns

Natural hazards: Cold periods in winter pose a threat to the unprepared.

Environment-current issues: air pollution from manufacturing and power plants contributing to acid rain; water pollution from industrial wastes, agricultural chemicals; habitat loss threatens wildlife populations

Environment-International Agreements

Party to: Air Pollution, Air Pollution-Nitrogen Oxides, Air Pollution-Sulphur 85, Air Pollution-Sulphur 94, Air Pollution-Volatile Organic Compounds, Antarctic-Environmental Protocol, Antarctic Treaty, Biodiversity, Climate Change, Desertification, Endangered Species, Environmental Modification, Hazardous Wastes, Law of the Sea, Marine Dumping, Marine Life Conservation, Nuclear Test Ban, Ozone Layer Protection. Ship Pollution, Tropical Timber 83, Tropical Timber 94, Wetlands, Whaling, Air Pollution–Persistent Organic Pollutants (signed 2001, ratified 2002), Climate Change–Kyoto Protocol (signed May 1998, ratified together with 14 other EU countries May 31 2002).

Geography-note: long boundary with Russia; Helsinki is northernmost national capital on European continent; population concentrated on small southwestern coastal plain

Geography of Iceland

Iceland is the second-largest island in Europe. The island is located east of Greenland and immediately south of the Arctic Circle, atop the divergent boundary of the Mid-Atlantic Ridge in the North Atlantic Ocean. It lies about 4,200 km (2,610 mi) from New York City and 860 km (534 mi) from Scotland.

Iceland has extensive volcanic and geothermal activity. The rift associated with the Mid-Atlantic Ridge, which marks the division between the European and North American tectonic plates, runs across Iceland from the southwest to the northeast. This geographic feature is prominent at the Pingvellir National Park, where the promontory creates an extraordinary natural amphitheatre. The site was the home of Iceland's parliament, the Alþing, which was first convened in 930.

About half of Iceland's land area, which is of recent volcanic origin, consists of a mountainous lava desert (highest elevation 2,119 m (6,952 ft) above sea level) and other wasteland. Eleven percent is covered by three large glaciers:

- Vatnajokull (8300 km^2)
- Langjokull (953 km^2)
- Hofsjokull (925 km^2).

And several smaller ones:

- Myrdalsjokull (695 km^2)
- Drangajokull (199 km^2)
- Eyjafjallajokull (107 km^2)
- Snæfellsjokull.

Twenty percent of the land is used for grazing, and only 1% is cultivated. An ambitious reforestation program is under way. Fossilized tree pollen and descriptions by the early settlers indicate that prior to human settlement in the 8-900s, trees covered about 30-40% of the island. Today, however, there are only small patches of the original birch forests left, the most prominent are Hallormsstaoaskogur and Vaglaskogur.

The inhabited areas are on the coast, particularly in the southwest; the central highlands are totally uninhabited.

Because of the Gulf Stream's moderating influence, the climate is characterized by damp, cool summers and relatively mild but windy winters. In Reykjavik, the average temperature is 11 °C (51.8 °F) in July and 0 °C (32 °F) in January (Koppen: Cfc).

Statistics

Geographic coordinates: Northern Europe (for cultural and historical reasons it is not considered to be a part of North America), between the Greenland Sea and the North Atlantic Ocean, northwest of the British Isles.

- North: Rifstangi, 66°322 3" N (Kolbeinsey, 67°08,9 N)
- South: Kötlutangi, 63°232 6" N (Surtsey, 63°17,7 N)
- West: Bjargtangar, 24°322 1" W
- East: Gerpir, 13°292 6" W (Hvalbakur, 13°16,6 W).

Area;

- Total: 103,125 km²
- Land: 100,329 km²
- Water: 2,796 km².

Area—comparative: Slightly smaller than Kentucky; about half the size of Great Britain; slightly larger than Hokkaido.

Land boundaries: 0 km

Coastline: 4,988 km

Maritime claims:

- Continental shelf: 200 nmi (370.4 km; 230.2 mi) or to the edge of the continental margin
- Exclusive economic zone: 200 nmi (370.4 km; 230.2 mi)
- Territorial sea: 12 nmi (22.2 km; 13.8 mi).

Climate: Temperate; moderated by North Atlantic Current; mild, windy winters; cool summers, damp in the South and West.

Terrain: Mostly plateau interspersed with mountain peaks, ice fields; coast deeply indented by bays and fjords.

Elevation extremes;

- Lowest point: Jokulsarlon Lagoon:-146 m, Atlantic Ocean 0 m
- Highest point: Hvannadalshnukur 2,110 m

Natural resources: Fish, hydropower, geothermal power.

Land use:

- Arable land: 0.07%
- Permanent crops: 0%
- Permanent pastures: 23%
- Forests and woodland: 1%
- Other: 76% (1993 est.)

Irrigated land: NA

Natural hazards: Earthquakes, volcanic activity, avalanches, and glacial lake outburst flooding (or *jokulhlaups*).

Environment—current issues: Water pollution from fertilizer runoff; inadequate wastewater treatment.

Environment—international agreements:

Party to: Air Pollution, Biodiversity, Climate Change, Desertification, Endangered Species, Hazardous Wastes, Law of the Sea, Marine Dumping, Nuclear Test Ban, Ozone Layer Protection, Ship Pollution (MARPOL 73/78), Wetlands, Whaling

Signed, but not ratified: Air Pollution-Persistent Organic Pollutants, Environmental Modification, Marine Life Conservation

Geography—note: Strategic location between Greenland and Europe; westernmost European country; more land covered by glaciers than in all of continental Europe.

Geography of Sweden

Sweden is a country in Northern Europe on the Scandinavian Peninsula. Its borders are Norway to the west; Finland to the northeast; and the Baltic Sea and Gulf of Bothnia to the east and south.

Sweden has a long coastline on the eastern side and the Scandinavian mountain chain (Skanderna) on the western border, a range that separates Sweden from Norway. It has maritime borders with Denmark, Germany, Poland, Russia, Lithuania, Latvia, and Estonia, and it is also linked to Denmark (southwest) by the Oresund Bridge. At 449,964 km^2 (173,732 sq mi), Sweden is the 55th largest country in the world. It is the 5th largest in Europe, and the largest in Northern Europe.

Terrain

Much of Sweden is heavily forested, with 78% of the country being forest and woodland, while farmland constitutes only 1% of land use. Southern Sweden is predominantly agricultural, with increasing forest coverage northward. Mountains and hills are dominant in the west. About 15% of Sweden lies north of the Arctic Circle.

The lowest elevation in Sweden is in the bay of Lake Hammarsjon, near Kristianstad at -2.41 m ("7.91 ft) below sea level. The highest point is Kebnekaise at 2,111 m (6,926 ft) above sea level.

The highest population density is in the Oresund region in southern Sweden and in the valley of lake Malaren near to Stockholm. Gotland and Oland are Sweden's largest islands; Vanern and Vattern are its largest lakes. Vanern is the third largest in Europe, after Lake Ladoga and Lake Onega in Russia.

Sweden consists of 39,030 km^2 of water area, constituting around 95,700 lakes. The lakes are sometimes used for water power plants, especially the large northern rivers and lakes. The two largest islands are Gotland and Oland in the southeast. They each have their own culture, most notably Gotland with the old, largely intact and heritage-filled city Visby.

Political Divisions

Provinces

Sweden has 25 provinces or *landskap* (landscapes), based on culture, geography and history: Bohuslan, Blekinge, Dalarna, Dalsland, Gotland, Gastrikland, Halland, Hälsingland, Harjedalen, Jamtland, Lapland, Medelpad, Norrbotten, Narke, Skane, Smaland, Sodermanland, Uppland, Varmland, Vastmanland, Vasterbotten, Vastergotland, Angermanland, Oland and Ostergotland.

While these provinces serve no political or administrative purpose, they play an important role for people's self-identification. The provinces are usually grouped together in three large lands (*landsdelar*): the northern Norrland, the central Svealand and southern Gotaland. The sparsely populated Norrland encompasses almost 60% of the country.

Counties

In each county there is also a separate County Council or *landsting*, which is the municipal representation appointed by the county electorate.

- K = Blekinge County
- W = Dalarna County
- I = Gotland County
- X = Gavleborg County
- N = Halland County
- Z = Jamtland County
- F = Jonkoping County
- H = Kalmar County
- G = Kronoberg County
- BD = Norrbotten County
- M = Skane County
- AB = Stockholm County
- D = Sodermanland County
- C = Uppsala County
- S = Varmland County
- AC = Vasterbotten County
- Y = Vasternorrland County
- U = Vastmanland County
- O = Vastra Gotaland County
- T = Orebro County
- E = Ostergotland County

The letters shown were on the vehicle registration plates until 1973.

Municipalities

Each county is further divided into municipalities or *kommuner*, ranging from only one (in Gotland County) to forty-nine (in Vastra Gotaland County): The total number of municipalities is 290.

The northern municipalities are often large in size, but have small populations – the largest municipality is Kiruna with an area as large as the three southern provinces in Sweden

(Scania, Blekinge and Halland) combined, but it only has a population of 25,000, and its density is about 1/km^2.

Population

Sweden has a population of about 9 million. The north is less populated than the southern and central parts, mostly because of its colder climate.

Cities

Cities and towns in Sweden are not political or administrative entities but localities or urban areas, independent of the municipal subdivision. The largest city, in terms of population, is the capital Stockholm, in the east, the dominant city for culture and media, with a population of 1,250,000. The second largest city is Gothenburg, with 510,500, in the west. The third largest is Malmo in the south, with 258,000. The largest city in the north is Umea with 76,000 inhabitants.

Natural Resources

Sweden's natural resources include copper, hydropower, iron ore, lead, silver, timber, uranium, and zinc.

Environment

Acid rain has become an issue because it is damaging soils and lakes and polluting the North Sea and the Baltic Sea. The HBV hydrology transport model has been used to analyze nutrient discharge to the Baltic from tributary watersheds.

Languages in Scandinavia

There are two language groups that have coexisted on the Scandinavian peninsula since prehistory—the North Germanic languages (Scandinavian languages) and the Sami languages. The majority languages on the peninsula, Swedish and Norwegian, are today, along with Danish, classified as Continental Scandinavian.

The North Germanic languages of Scandinavia are traditionally divided into an East Scandinavian branch (Danish and Swedish) and a West Scandinavian branch (Norwegian, Icelandic, and Faroese), but because of changes appearing in the languages since 1600, the East Scandinavian and West Scandinavian branches are now usually reconfigured into

Insular Scandinavian (*o-nordisk/oy-nordisk*) featuring (Icelandic and Faroese) and Continental Scandinavian (*Skandinavisk*), comprising Danish, Norwegian, and Swedish. The modern division is based on the degree of mutual comprehensibility between the languages in the two branches. Note that *skandinavisk(a)* may also refer to a way of speaking one Scandinavian language in a way intended to be more easily understood by speakers of the other Scandinavian languages.

Apart from Sami and the languages of minority groups speaking a variant of the majority language of a neighboring state, the following minority languages in Scandinavia are protected under the European Charter for Regional or Minority Languages: Yiddish, Romani Chib, Romanes and Romani.

Continental Scandinavian Languages

The dialects of Denmark, Norway and Sweden form a dialect continuum and are mutually intelligible. The populations of the Scandinavian countries, with a Scandinavian mother tongue, can—at least with some training—understand each other's standard languages as they appear in print and are heard on radio and television. The reason why Danish, Swedish and the two official written versions of Norwegian (*Nynorsk* and *Bokmal*) are traditionally viewed as different languages, rather than dialects of one common language, is that they each are well established standard languages in their respective countries. They are related to, but not mutually intelligible with, the other North Germanic languages, Icelandic and Faroese, which are descended from Old West Norse. Danish, Swedish and Norwegian have, since medieval times, been influenced to varying degrees by Middle Low German and standard German. A substantial amount of that influence was a by-product of the economic activity generated by the Hanseatic League.

Norwegians are accustomed to variation, and may perceive Danish and Swedish only as slightly more distant dialects. This is because they have two official written standards, in addition to the habit of strongly holding on to local dialects. The people of Stockholm, Sweden and Copenhagen, Denmark, have the greatest difficulty in understanding other Scandinavian languages. In the Faroe Islands, learning Danish is mandatory. This causes Faroese people to become bilingual in two very distinct North Germanic languages, making it relatively easy for them to understand the other two Mainland Scandinavian languages.

The Scandinavian languages are (as a language family) entirely unrelated to Finnish, Estonian, and Sami languages, which as Finno-Ugric languages are distantly related to Hungarian. Owing to the close proximity, there is still a great deal of borrowing from the Swedish and Norwegian languages in the Finnish, Estonian, and Sami languages. The long history of linguistic influence of Swedish on Finnish is also due to the fact that Finnish, the language of the majority in Finland, was treated as a minority language while Finland was part of Sweden. Finnish-speakers had to learn Swedish in order to advance to higher positions. Although Iceland was under the political control of Denmark until a much later date (1918), very little influence and borrowing from Danish has occurred in the Icelandic language. Icelandic remained the preferred language among the ruling classes in Iceland; Danish was not used for official communications, most of the royal officials were of Icelandic descent and the language of the church and law courts remained Icelandic.

Sami Languages

The Sami languages are indigenous minority languages in Scandinavia. They belong to the Finno-Ugric branch of the Uralic language family and are unrelated to the North Germanic languages other than by limited grammatical (particularly lexical) characteristics resulting from prolonged contact. Sami is divided into several languages or dialects Consonant gradation is a feature in both Finnish and northern Sami dialects, but it is not present in south Sami, which is considered to have a different language history. According to the Sami Information Centre of the Sami Parliament in Sweden, southern Sami originated in an earlier migration from the south into the Scandinavian peninsula.

Finland and Scandinavia

Finland is officially bilingual, with Finnish and Swedish having mostly the same status at national level. Native Swedish speakers constitute a small, but rather influential, minority. Children are taught the other official language at school; for Swedish-speakers, this is Finnish (usually from the 3rd grade), and for Finnish-speakers, Swedish (usually from the 7th grade). Finnish speakers constitute a language minority in Sweden, Norway and Russian Federation. There are also Finnic languages different from standard Finnish, known as Meankieli in Sweden and Kven in Norway.

History

During a period of Christianization and state formation in the 10th-13th centuries, three consolidated kingdoms emerged in Scandinavia:

- Denmark, forged from the Lands of Denmark (including Jutland, Zealand and Scania (Skaneland) on the Scandinavian Peninsula. The island Gotland in modern-day Sweden was initially also part of the Danish realm.)
- Sweden, forged from the Lands of Sweden on the Scandinavian Peninsula (excluding the provinces Bohuslan, Harjedalen, Jamtland and Idre & Sarna, Halland, Blekinge and Scania of modern-day Sweden)
- Norway (including Bohuslan, Harjedalen, Jamtland and Idre & Sarna on the Scandinavian Peninsula, and its island colonies Iceland, Greenland, Faroe Islands, Shetland, Orkney, Isle of Man and the Hebrides.)

In the 1645 Treaty of Bromsebro, Denmark–Norway ceded the Norwegian provinces of Jamtland, Harjedalen and Idre & Sarna, as well as the Baltic Sea islands of Gotland and Osel (in Estonia) to Sweden. The Treaty of Roskilde, signed in 1658, forced Denmark–Norway to cede the Danish provinces Scania, Blekinge, Halland, Bornholm and the Norwegian provinces of Bahuslen and Trondelag to Sweden. The 1660 Treaty of Copenhagen forced Sweden to return Bornholm and Trondelag to Denmark–Norway, and to give up its recent claims to the island Funen.

Scandinavian Unions

The three Scandinavian kingdoms were united in 1397 in the Kalmar Union by Queen Margaret I of Denmark. Sweden left the union in 1523 under King Gustav Vasa. In the aftermath of Sweden's secession from the Kalmar Union, civil war broke out in Denmark and Norway. The Protestant Reformation followed. When things had settled down, the Norwegian Privy

Council was abolished—it assembled for the last time in 1537. A personal union, entered into by the kingdoms of Denmark and Norway in 1536, lasted until 1814. Three sovereign successor states have subsequently emerged from this unequal union: Denmark, Norway and Iceland.

Denmark–Norway is the historiographical name for the former political union consisting of the kingdoms of Denmark and Norway, including the Norwegian dependencies of Iceland, Greenland and the Faroe Islands. The corresponding adjective and demonym is Dano-Norwegian. During Danish rule, Norway kept its separate laws, coinage and army, as well as some institutions such as a royal chancellor. Norway's old royal line had died out with the death of Olav IV, but Norway's remaining a hereditary kingdom was an important factor to the Oldenburg dynasty of Denmark–Norway in its struggles to win elections as kings of Denmark.

The Dano-Norwegian union was formally dissolved at the 1814 Treaty of Kiel. The territory of Norway proper was ceded to the King of Sweden, but Norway's overseas possessions were kept by Denmark. However, widespread Norwegian resistance to the prospect of a union with Sweden induced the governor of Norway, crown prince Christian Frederick (later Christian VIII of Denmark), to call a constituent assembly at Eidsvoll in April 1814. The assembly drew up a liberal constitution and elected him to the throne of Norway. Following a Swedish invasion during the summer, the peace conditions specified that king Christian Frederik had to resign, but Norway was to keep its independence and its constitution within a personal union with Sweden. Christian Frederik formally abdicated on August 10, 1814 and returned to Denmark. The parliament Storting elected king Charles XIII of Sweden as king of Norway on November 4.

The union between Sweden and Norway was dissolved in 1905, after which Prince Charles of Denmark was elected king of Norway under the name of Haakon VII.

Political Scandinavism

The modern usage of the term *Scandinavia* has been influenced by Scandinavism (the Scandinavist political movement), which was active in the middle of the 19th century, mainly between the First war of Schleswig (1848–1850), in which Sweden and Norway contributed with considerable military force, and the Second war of Schleswig (1864). In 1864, the Swedish parliament denounced the promises of military support made to Denmark by Charles XV of Sweden. The members of the Swedish parliament were wary of joining an alliance against the rising German power.

The Swedish king also proposed a unification of Denmark, Norway and Sweden into a single united kingdom. The background for the proposal was the tumultuous events during the Napoleonic wars in the beginning of the century. This war resulted in Finland (formerly the eastern third of Sweden) becoming the Russian Grand Duchy of Finland in 1809 and Norway (*de jure* in union with Denmark since 1387, although *de facto* treated as a province) becoming independent in 1814, but thereafter swiftly forced to accept a personal union with Sweden. The dependent territories Iceland, the Faroe Islands and Greenland, historically part of Norway, remained with Denmark in accordance with the Treaty of Kiel. Sweden and Norway were thus united under the Swedish monarch, but Finland's inclusion in the Russian Empire excluded any possibility for a political union between Finland and any of the other Nordic countries.

The end of the Scandinavian political movement came when Denmark was denied the military support promised from Sweden and Norway to annex the (Danish) Duchy of Schleswig, which together with the (German) Duchy of Holstein had been in personal union with Denmark. The Second war of Schleswig followed in 1864, a brief but disastrous war between Denmark and Prussia (supported by Austria). Schleswig-Holstein was conquered by Prussia, and after Prussia's success in the Franco-Prussian War a Prussian-led German Empire was created, and a new power-balance of the Baltic sea countries was established.

Even if a Scandinavian political union never came about at this point, there was a Scandinavian Monetary Union established in 1873, lasting until World War I.

Finland

Finland, officially the Republic of Finland, is a Nordic country situated in the Fennoscandian region of Northern

Europe. It is bordered by Sweden on the west, Norway on the north and Russia on the east, while Estonia lies to its south across the Gulf of Finland.

Around 5.4 million people reside in Finland, with the majority concentrated in the southern part of the country. It is the eighth largest country in Europe in terms of area and the most sparsely populated country in the European Union. Finland is a parliamentary republic with a central government based in Helsinki and local governments in 342 municipalities. A total of about one million residents live in the Greater Helsinki area (which includes Helsinki, Espoo, Kauniainen and Vantaa), and a third of the country's GDP is produced there. Other larger cities include Tampere, Turku, Oulu, Jyvaskyla, Kuopio and Lahti.

Finland was historically a part of Sweden and from 1809 an autonomous Grand Duchy within the Russian Empire. Finland's declaration of independence from Russia in 1917 was followed by a civil war, wars against the Soviet Union and Nazi Germany, and a period of official neutrality during the Cold War. Finland joined the United Nations in 1955, the OECD in 1969, the European Union in 1995, and the eurozone since its beginning. Finland has been ranked the second most stable country in the world, in a survey based on social, economic, political and military indicators.

Finland was a relative latecomer to industrialization, remaining a largely agrarian country until the 1950s. Thereafter, economic development was rapid, and the country reached the world's top income levels in the 1970s. Between 1970 and 1990, Finland built an extensive welfare state. In the aftermath of the country's severe depression in the early 1990s, successive governments have changed the Finnish economic system through some privatisation, deregulation and tax cuts.

Finland is well placed in international comparisons of national performance such as the share of high-technology manufacturing and health care. The country is ranked 1st in the 2009 Legatum Prosperity rating, which is based on economical performance and quality of life.

Etymology

The name *Suomi* (Finnish for "Finland") has uncertain

origins, but a candidate for a cognate is the Proto-Baltic word **zeme,* meaning "land". In addition to the close relatives of Finnish (the Baltic-Finnic languages), this name is also used in the Baltic languages Latvian and Lithuanian. According to an earlier theory the name was derived from *suomaa* (fen land) or *suoniemi* (fen cape).

The Swedish-language name *Finland* has resemblance with the North Germanic placenames Finnmark, Finnveden and hundreds of other toponyms starting with *Fin(n)* in Sweden and Norway. Some of these names are obviously derived from *finnr*, a Germanic word for a wanderer/finder and thus supposedly meaning nomadic "hunter-gatherers" or slash and burn agriculturists as opposed to the Germanic sedentary farmers and seafaring traders and pirates. The term "Finn" often refers to Sami people, too. *Finn* was used to refer to the people of Finland Proper after the 15th century, when the church appointed a bishop — who became one of the most powerful men in the province — over the whole area corresponding roughly to today's Finland. The fact that there was no other ecclesiastical authority of the same level, coupled with the bishop's temporal authority, engendered a sense of "the Finns" belonging to one geographical area over which the name spread from the 15th century onwards to refer to the people of the entire country.

Among the first documents to mention "a land of the Finns" are two rune-stones. There is one in Soderby, Sweden, with the inscription *finlont* (U 582) and one in Gotland, a Swedish island in the Baltic Sea, with the inscription *finlandi* (G 319), dating from the 11th century.

History

Prehistory

According to archaeological evidence, the area now comprising Finland was settled at the latest around 8500 BCE during the Stone Age as the ice sheet of the last ice age receded. The artifacts the first settlers left behind present characteristics that are shared with those found in Estonia, Russia and Norway. The earliest people were hunter-gatherers, using stone tools. There is also evidence of carved stone animal heads. The first pottery appeared in 3000 BCE when settlers from the East brought in the Comb Ceramic culture. The arrival of the Corded Ware culture in southern coastal Finland between 3000–2500 BCE coincided with the start of agriculture. Even with the introduction of agriculture, hunting and fishing continued to be important parts of the subsistence economy.

The Bronze Age (1500–500 BCE) and Iron Age (500 BCE–1200 CE) were characterised by extensive contacts with other cultures in the Fennoscandian and Baltic regions. There is no consensus on when Finno-Ugric languages and Indo-European languages were first spoken in the area of contemporary Finland.

In the beginning of 9th century the inhabited area of Finnish tribes was bordered by:

- White Sea, in the north
- Northmen, in the west behind Scandinavian Mountains
- Swedes & Goths, in the southern tip of Scandinavian Peninsula
- Baltic Sea
- Estland, to the south of Gulf of Finland
- Slavonic tribes, in the Novgorod surroundings and further in the south
- Kingdom of Chazars, Morduines, White Bulgarians and Bashkirs who lived in the Volga watershed area.

Politics and Government

The Constitution of Finland defines the political system. Finland is a representative democracy with a semi-presidential parliamentary system. Aside from state-level politics, residents use their vote in municipal elections and in the European Union elections.

According to the Constitution, the President of Finland is the head of state and responsible for foreign policy (which excludes affairs related to the European Union) in cooperation with the cabinet. Other powers include Commander-in-Chief, decree, and appointive powers. Direct vote is used to elect the president for a term of six years and maximum two consecutive terms. The current president is Tarja Halonen (SDP).

The 200-member unicameral Parliament of Finland exercises the supreme legislative authority in Finland. The parliament may alter laws and the constitution, bring about the resignation of the Council of State, and override presidential vetoes. Its acts are not subject to judicial review. Various parliament committees listen to experts and prepare legislation. Proportional vote in multi-seat constituencies is used to elect the parliament for a term of four years. The Speaker of Parliament is currently Sauli Niinisto (National Coalition Party). The cabinet (the Finnish Council of State) exercises most executive powers. It is headed by the Prime Minister of Finland and includes other ministers and the Chancellor of Justice. Parliament majority decides its composition, and a vote of no confidence can be used to modify it. The current prime minister is Mari Kiviniemi (Centre Party).

Since equal and common suffrage was introduced in 1906, the parliament has been dominated by the Centre Party (former Agrarian Union), National Coalition Party and Social Democrats, which have approximately equal support and represent 65–80% of voters. After 1944 Communists were a factor to consider for a few decades. The relative strengths of the parties vary only slightly in the elections because of the proportional election from multi-member districts, but there are some visible long-term trends. The autonomous Aland islands has separate elections, where Liberals for Aland was the largest party in 2007 elections.

Economy

Finland has a highly industrialized free-market economy with a per capita output equal to that of other European economies such as France, Germany, Belgium or the UK. The largest sector of the economy is services at 65.7%, followed by manufacturing and refining at 31.4%. Primary production is 2.9%. With respect to foreign trade, the key economic sector is manufacturing. The largest industries are electronics (21.6%), machinery, vehicles and other engineered metal products (21.1%), forest industry (13.1%) and chemicals (10.9%).

Finland has timber and several mineral and freshwater resources. Forestry, paper factories, and the agricultural sector (on which taxpayers spend around 3 billion euros annually) are politically sensitive to rural residents. The Greater Helsinki area generates around a third of GDP. In a 2004 OECD comparison, high-technology manufacturing in Finland ranked second largest after Ireland. Knowledge-intensive services have also ranked the smallest and slow-growth sectors – especially agriculture and low-technology manufacturing – second largest after Ireland. Overall short-term outlook was good, and GDP growth has been above many EU peers.

Finland is highly integrated in the global economy, and international trade is a third of GDP. The European Union makes 60% of the total trade. The largest trade flows are with Germany, Russia, Sweden, United Kingdom, United States, Netherlands and China. Trade policy is managed by the European Union, where Finland has traditionally been among the free trade supporters, except for agriculture. Finland is the only Nordic country to have joined the Eurozone.

Private sector employees amount to 1.8 million, out of which around a third with tertiary education. The average cost of a private sector employee per hour was 25.1 euros in 2004. As of 2008 average purchasing power-adjusted income levels are similar to those of Italy, Sweden, Germany and France. In 2006, 62% of the workforce worked for enterprises with less than 250 employees and they accounted for 49% of total business turnover and had the strongest rate of growth. The female employment rate is high. Gender segregation between male-dominated professions and female-dominated professions is higher than in the US. The proportion of part-time workers was one of the lowest in OECD in 1999.

Employment rate 68% and unemployment rate was 6.8% in early 2008. 18% of residents are outside job market at the age of 50 and less than a third working at the age of 61. Unfunded pensions and other promises such as health insurances are a dominant future liability, though Finland is much better prepared than countries such as France or Germany. Directly held public debt has been reduced to around 32% of GDP in 2007. In 2007, the average household savings rate was-3.8 and household debt 101% of annual disposable income, a typical level in Europe. Home ownership rate is 60%.

As of 2006, 2.4 million households reside in Finland. The average size is 2.1 persons; 40% of households consist of a single person, 32% two persons and 28% three or more persons. Residential buildings total 1.2 million and the average residential space is 38 m per person. The average residential property without land costs 1,187 euro per sq metre and residential land 8.6 euro per sq metre. 74% of households had a car. There are 2.5 million cars and 0.4 million other vehicles.

Around 92% have a mobile phone and 83.5% (2009) Internet connection at home. The average total household consumption was 20,000 euro, out of which housing consisted of about 5500 euro, transport about 3000 euro, food and beverages excluding alcoholic at around 2500 euro, recreation and culture at around 2000 euro. Purchasing power-adjusted average household consumption is about the same level as it is in Germany, Sweden and Italy. According to Invest in Finland, private consumption grew by 3% in 2006 and consumer trends included durables, high quality products, and spending on well-being.

Education and Science

Most pre-tertiary education is arranged at municipal level. Even though many or most schools were started as private schools, today only around 3% students are enrolled in private schools (mostly Helsinki-based schools such as SYK), many times less than in Sweden and most other developed countries. Pre-school education is rare compared to other EU countries. Formal education is usually started at the age of 7. The primary school takes normally 6 years, the lower secondary school 3 years, and most schools are managed by municipal officials.

The flexible curriculum is set by the Ministry of Education and the Education Board. Attendance is compulsory between the ages of 7 and 16. After lower secondary school, graduates may either enter the workforce directly, or apply to trade schools or gymnasiums (upper secondary schools). Trade schools prepare for professions. Academically oriented gymnasiums have higher entrance requirements and specifically prepare for Abitur and tertiary education. Graduation from either formally qualifies for tertiary education.

In tertiary education, two mostly separate and non-interoperating sectors are found: the profession-oriented polytechnics and the research-oriented universities. Finns used to take student loans and scholarships, but for the past decades the financial risk has been moved solely to the government. There are 20 universities and 30 polytechnics in the country. Helsinki University is 108 in the Top University Ranking of

2009. The World Economic Forum ranks Finland's tertiary education #1 in the world. Around 33% of residents have a tertiary degree, similar to Nordics and more than in most other OECD countries except Canada (44%), United States (38%) and Japan(37%). The proportion of foreign students is 3% of all tertiary enrolments, one of the lowest in OECD, while in advanced programs it is 7.3%, still below OECD average 16.5%.

More than 30% of tertiary graduates are in science-related fields. Finnish researchers are leading contributors to such fields as forest improvement, new materials, the environment, neural networks, low-temperature physics, brain research, biotechnology, genetic technology and communications.

Finland is highly productive in scientific research. In 2005, Finland had the fourth most scientific publications per capita of the OECD countries. In 2007, 1801 patents were filed in Finland.

Energy

Anyone can enter the free and largely privately owned financial and physical Nordic energy markets traded in Nord Pool and Nord Pool Spot exchanges, which have provided competitive prices compared to other EU countries. As of 2007, Finland has roughly the lowest industrial electricity prices in the EU-15 (equal to France).

In 2006, the energy market was around 90 terawatt hours and the peak demand around 15 gigawatts in winter. This means that the energy consumption per capita is around 7.2 tons of oil equivalent per year. Industry and construction consumed 51% of total consumption, a relatively high figure reflecting Finland's industries. Finland's hydrocarbon resources are limited to peat and wood. Finland has little hydropower capacity compared to Sweden or Norway. Most energy demand is satisfied with fossil fuels such as coal, oil and natural gas.

Finland has four privately owned nuclear reactors producing 18% of the country's energy, one research reactor in Otaniemi campus, and the fifth AREVA-Siemens-built reactor – the world's largest at 1600 MWe and a focal point of Europe's nuclear industry – is scheduled to be operational by 2011. Renewable energy forms (industry-burned wood, consumer-burned wood, peat, industrial residue, garbage) make high 25%

compared to the EU average 10%. A varying amount (5–17%) of electricity has been imported from Russia (at around 3 gigawatt power line capacity), Sweden and Norway.

A new submarine power cable from Russia has been considered a national security issue, and one permit application has already been rejected. Finland negotiated itself expensive Kyoto and EU emission terms. They are causing a sharp increase in energy prices and 1-2 billion euro annual cost, amplified by the aging and soon decommissioned production capacity. Energy companies are ready to increase nuclear power production, if parliament granted permits for new reactors.

Transportation

The extensive road system is utilized by most internal cargo and passenger traffic. The annual road network expenditure of around 1 billion euro is paid with vehicle and fuel taxes which amount to around 1.5 billion euro and 1 billion euro.

The main international passenger gateway is Helsinki-Vantaa Airport with over 13 million passengers in 2008. Oulu Airport is the second largest and around 25 airports have scheduled passenger services. The Helsinki-Vantaa based Finnair, Blue and Finncomm Airlines sell air services both domestically and internationally. Helsinki has an optimal location for great circle routes between Western Europe and the Far East. Despite low population density, the Government spends annually around 350 million euro in maintaining 5,865 kilometres (3,644 mi) railway tracks even to many rural towns. Only one rail company operates in Finland, VR Group, which has 5% passenger market share (out of which 80% are urban trips in Greater Helsinki) and 25% cargo market share. Helsinki has an urban rail network.

The majority of international cargo utilizes ports. Port logistics prices are low. Vuosaari harbour in Helsinki is the largest container port after completion in 2008 and others include Hamina, Hanko, Pori, Rauma, Oulu. There is passenger traffic from Helsinki and Turku, which have ferry connections to Tallinn, Mariehamn and Stockholm. The Helsinki–Tallinn route, one of the busiest passenger sea routes in the world, has also been served by a helicopter line.

Industry

Finland has developed greatly since 1945, when it was a primarily agricultural nation, and created major firms in telecommunications like Nokia, electronics, metalworking, forestry, metrology and climate measurement systems like Vaisala, and construction like Pöyry. Shipbuilding industry is important for the Finnish economy, and the world's biggest cruise ships are built in Finnish shipyards.

Public Policy

Finnish politicians have often emulated other Nordics and the Nordic model. Nordics have been free-trading and relatively welcoming to skilled migrants for over a century, though in Finland immigration is relatively new. The level of protection in commodity trade has been low, except for agricultural products.

Finland has top levels of economic freedom in many areas, although there is a heavy tax burden and inflexible job market. Finland is ranked 16th (ninth in Europe) in the 2008 Index of Economic Freedom. Recently, Finland has topped the patents per capita statistics, and overall productivity growth has been strong in areas such as electronics. While the manufacturing sector is thriving, OECD points out that the service sector would benefit substantially from policy improvements. Finland is one of the most fiscally responsible EU countries. IMD World Competitiveness Yearbook 2007 ranked Finland 17th most competitive. The World Economic Forum 2008 index ranked Finland the 6th most competitive. In both indicators, Finland's performance was next to Germany, and significantly higher than most European countries. In the Business competitiveness index 2007-08 Finland ranked third in the world.

Economists attribute much growth to reforms in the product markets. According to OECD, only four EU-15 countries have less regulated product markets (UK, Ireland, Denmark and Sweden) and only one has less regulated financial markets (Denmark). Nordic countries were pioneers in liberalizing energy, postal, and other markets in Europe. The legal system is clear and business bureaucracy less than most countries. Property rights are well protected and contractual agreements are strictly honoured. Finland is rated the 6th least corrupted countries in Corruption perception index. Finland is rated 13th in the Ease of Doing Business Index. It indicates exceptional ease to trade across borders (5th), enforce contracts (7th), and close a business (5th), and exceptional hardship to employ workers (127th) and pay taxes (83rd).

Finnish job market regulation is a remaining example of Nordic neocorporatist model. In the 1990s, Denmark liberalized its job market, Sweden moved to more flexible decentralized contracts, and Finnish trade unions blocked most reforms. Finnish law forces all workers to obey the national contracts that are drafted every few years for each profession and seniority level. The agreement becomes universally enforceable provided that more than 50% of the employees support it, in practice by being a member of a relevant trade union. The unionization rate is high (70%), especially in the middle class (AKAVA – 80%). A lack of a national agreement in an industry is considered an exception. More flexibility is generally recommended by economists for various reasons.

Numismatics

In Finland, the euro was introduced in 2002. As a preparation for this date, the minting of the new euro coins started as early as 1999; this is why the first euro coins from Finland have the year 1999 on them, instead of 2002 like some of the other countries of the Eurozone. Three different designs (one for €2 coin, one for €1 coin and one for the other six coins) were selected for the Finnish coins. In 2007, in order to adopt the new common map like the rest of the Eurozone countries, Finland changed the common side of their coins.

Finland also has a rich collection of collectors' coins, with face value ranging from 5 to 100 euro. These coins are a legacy of an old national practice of minting silver and gold commemorative coins. Unlike normal issues, these coins are not legal tender in all the eurozone; for instance, a €5 Finnish commemorative coin cannot be used in any other country.

Tourism

In 2005, Finnish tourism grossed over €6.7 billion with a five percent increase from the previous year. Much of the sudden growth can be attributed to the globalisation and modernisation of the country as well as a rise in positive publicity and awareness. There are many attractions in Finland which attracted over 4 million visitors in 2005. The Finnish landscape is covered with thick pine forests, rolling hills and complemented with a labyrinth of lakes and inlets. Much of Finland is pristine and virgin as it contains 35 national parks from the Southern shores of the Gulf of Finland to the high fells of Lapland. It is also an urbanised region with many cultural events and activities. Commercial cruises between major coastal and port cities in the Baltic region, including Helsinki, Turku, Tallinn, Stockholm and Travemünde, play a significant role in the local tourism industry. Finland is regarded as the home of Saint Nicholas or Santa Claus, living in the northern Lapland region. Above the Arctic Circle, there is a polar night, a period when the sun does not rise for days or weeks, or even months. Lapland is so far north that the Aurora Borealis, atmospheric fluorescence, is seen regularly in winter.

Outdoor activities range from Nordic skiing, golf, fishing, yachting, lake cruises, hiking, kayaking among many others. At Finland's northernmost point, in the heart of summer, the Sun does not completely set for 73 consecutive days. Wildlife is abundant in Finland. Bird-watching is popular for those fond of flying fauna, however hunting is also popular. Elk, reindeer and hare are all common game in Finland. Olavinlinna in Savonlinna hosts the annual Savonlinna Opera Festival.

Culture

Literature

Though Finnish written language could be said to exist since Mikael Agricola translated the New Testament into Finnish in the sixteenth century as a result of the Protestant Reformation, few notable works of literature were written until the nineteenth century, which saw the beginning of a Finnish national Romantic Movement. This prompted Elias Lönnrot to collect Finnish and Karelian folk poetry and arrange and publish them as *Kalevala*, the Finnish national epic. The era saw a rise of poets and novelists who wrote in Finnish, notably Aleksis Kivi and Eino Leino.

After Finland became independent there was a rise of modernist writers, most famously Mika Waltari. Frans Eemil Sillanpaa was awarded the Nobel Prize in Literature in 1939.

The second World War prompted a return to more national interests in comparison to a more international line of thought, characterized by Vaino Linna. Literature in modern Finland is in a healthy state. Popular modern writers include Arto Paasilinna, Ilkka Remes, Kari Hotakainen and Jari Tervo, while the best novel is annually awarded the prestigious Finlandia Prize.

Visual Arts

Finns have made major contributions to handicrafts and industrial design. Finland's best-known sculptor of the twentieth century was Waino Aaltonen, remembered for his monumental busts and sculptures. Finnish architecture is famous around the world. Among the top of the twentieth century Finnish architects to win international recognition are Eliel Saarinen (designer of the widely recognised Helsinki Central railway station and many other public works) and his son Eero Saarinen. Alvar Aalto, who helped bring functionalist architecture to Finland, is also famous for his work in furniture, textiles and glassware.

Music

Folk and Sami music: Much of the music of Finland is influenced by traditional Karelian melodies and lyrics, as comprised in the *Kalevala*. Karelian culture is perceived as the purest expression of the Finnic myths and beliefs, less influenced by Germanic influence, in contrast to Finland's position between the East and the West. Finnish folk music has undergone a roots revival in recent decades, and has become a part of popular music.

The people of northern Finland, Sweden and Norway, the Sami, are known primarily for highly spiritual songs called Joik. The same word sometimes refers to lavlu or vuelie songs, though this is technically incorrect.

Classical and opera: The first Finnish opera was written by the German composer Fredrik Pacius in 1852. Pacius also wrote *Maamme / Vart land* (Our Country), Finland's national anthem. In the 1890s Finnish nationalism based on the *Kalevala* spread, and Jean Sibelius became famous for his vocal symphony *Kullervo*. He soon received a grant to study *runo singers* in Karelia and continued his rise as the first prominent Finnish musician. In 1899 he composed Finlandia, which played its important role in Finland gaining independence. He remains one of Finland's most popular national figures and is a symbol of the nation.

Today, Finland has a very lively classical music scene. Finnish classical music has only existed for about a hundred years, and many of the important composers are still alive, such as Magnus Lindberg, Kaija Saariaho, Aulis Sallinen and Einojuhani Rautavaara. The composers are accompanied with a large number of great conductors such as Sakari Oramo, Mikko Franck, Esa-Pekka Salonen, Osmo Vanska, Jukka-Pekka Saraste, Susanna Malkki and Leif Segerstam. Some of the internationally acclaimed Finnish classical musicians are Karita Mattila, Soile Isokoski, Kari Kriikku, Pekka Kuusisto, Reka Szilvay and Linda Brava.

Popular music: Modern Finnish popular music includes a number of prominent rock bands, jazz musicians, hip hop performers, and dance music acts such as HIM, Bomfunk MCs and Darude. Finnish electronic music such as the Sahko Recordings record label enjoys underground acclaim. *Iskelma* (coined directly from the German word *Schlager*, meaning *hit*) is a traditional Finnish word for a light popular song.

Finnish popular music also includes various kinds of dance music; tango, a style of Argentine music, is also popular. One of the most productive composers of popular music was Toivo Karki, and the most famous singer Olavi Virta (1915–1972). Among the lyricists, Sauvo Puhtila (born 1928), Reino Helismaa (1913–1965) and Veikko "Vexi" Salmi are the most remarkable authors. The composer and bandleader Jimi Tenor is well known for his brand of retro-funk music.

Dance and rock music: Notable Finnish dance and electronic music artists include Jori Hulkkonen, Darude, JS16, DJ Proteus and DJ Orkidea.

The Finnish rock-music scene emerged in 1960s with pioneers such as Blues Section and Kirka. In the 1970s Finnish rock musicians, such as Juice Leskinen, and a pop rock group called Kaseva started to write their own music instead of translating international hits into Finnish. During the decade some progressive rock groups, such as Tasavallan Presidentti and Wigwam, gained respect abroad but failed to make a commercial breakthrough outside Finland. This was also the fate of the rock and roll group Hurriganes. The Finnish punk scene produced some internationally acknowledged names including Terveet Kadet in 1980s. Hanoi Rocks was a pioneering 1980s glam rock act that left perhaps a deeper mark in the history of popular music than any other Finnish group, giving inspiration for Guns N' Roses. HIM is by far Finland's most internationally known band, going gold in the United States with the album Dark Light, and receiving a Grammy nomination in 2008. The band's frontman Ville Valo is one of Finland and Europe's most famous and intriguing entertainers. The band maintains a massive international cult following. Apocalyptica are an internationally famous Finnish group who are most renowned for mixing strings led classical music with classic heavy metal. Primarily the band found fame through covering songs from Metallica but branched out to writing and recording their own material which featured guest vocalists such as Cristina Scabbia of Lacuna Coil and Til Lindemann of Rammstein. When Finland hosted the Eurovision Song Contest in 2007, Apocalyptica was one of the Finnish bands selected to perform during the ceremony as part of a showcase of modern Finnish music. Of note is the increasingly higher number of metal bands, among which Ensiferum, Wintersun and Turisas, that have gained international recognition for their innovative style and quality. Also a very known melodic heavy metal band Nightwish comes from Finland. Another well known metal band is Children of Bodom.

Cinema: In film industry, notable directors include Aki Kaurismaki, Mauritz Stiller, Spede Pasanen and Hollywood film director and producer Renny Harlin.

Media and Communications

Today there are 200 newspapers; 320 popular magazines, 2,100 professional magazines and 67 commercial radio stations, with one nationwide, five national public service radio channels, three digital radio channels. Each year around twelve feature films are made, 12,000 book titles published and 12 million records sold.

Sanoma publishes the newspaper *Helsingin Sanomat* (the circulation of 412,000 making it the largest newspaper), the tabloid *Ilta-Sanomat,* the commerce-oriented *Taloussanomat,* and the television channel Nelonen. The other major publisher Alma Media publishes over thirty magazines, including newspaper *Aamulehti,* tabloid *Iltalehti* and commerce-oriented *Kauppalehti.* Finns, along with other Nordic people and the Japanese, spend the most time in the world reading newspapers.

The National Broadcasting Company YLE has five television channels and 13 radio channels in two national languages. YLE is funded through a mandatory license for television owners and fees for private broadcasters. All TV channels are broadcast digitally, both terrestrially and on cable. The most popular television channel MTV3 and the most popular radio channel Radio Nova are owned by Nordic Broadcasting (Bonnier and Proventus Industrier).

Around 79 percent of the population use the Internet. Finland had around 1.52 million broadband Internet connections by the end of June 2007 or around 287 per 1,000 inhabitants. All Finnish schools and public libraries have Internet connections and computers. Most residents have a mobile phone. It's used mostly for contact and value-added services are rare. In October 2009, Finland's Ministry of Transport and Communications committed to ensuring that every person in Finland will be able to access the internet at a minimum speed of one megabit-per-second beginning July 2010.

Public Holidays

All official holidays in Finland are established by acts of Parliament. The official holidays can be divided into Christian and secular holidays. The main Christian holidays are Christmas, Epiphany, Easter, Ascension Day, Pentecost and All Saints Day. The secular holidays are New Year's Day, May Day, Midsummer Day and the Independence Day. Christmas is the most extensively celebrated holiday: usually at least 23rd to 26 December are holidays. Also in the region of Bothnia (usually referred at the city of Kokkola) there is an celebration called Venetsialaiset, the celebration of water and fire.

Sports

Various sporting events are popular in Finland. Pesapallo (reminiscent of baseball) is the national sport of Finland, although the most popular sports in Finland in terms of media coverage are Formula One, rallying, ice hockey and football.

Finland won the ice-hockey world championship once in 1995. Jari Kurri and Teemu Selanne are the two Finnish-born ice hockey players to have scored 600 goals in their NHL careers.

The Finland national football team has never qualified for a finals tournament of the World Cup or the European Championships. Jari Litmanen, Sami Hyypia, Antti Niemi, Jussi Jaaskelainen and Mikael Forssell are the most internationally renowned of the Finnish football players. Snowboarding is also very popular in Finland, and there are many Finnish professional snowboarders such as Antti Autti, Heikki Sorsa, Jussi Oksanen, Eero Ettala, Peetu Piiroinen and Joni Malmi.

Relative to its population, Finland has been a top country in the world in automobile racing, measured by international success. Finland has produced three Formula One World Champions – Keke Rosberg (Williams, 1982), Mika Hakkinen (McLaren, 1998 and 1999) and Kimi Raikkonen (Ferrari, 2007). Following Raikkonen's departure from the sport, the only Finnish Formula One driver currently active is Heikki Kovalainen (Lotus). Rosberg's son, Nico Rosberg (Mercedes GP), is also currently driving, but under his mother's German nationality.

Other notable Finnish Grand Prix drivers include Leo Kinnunen, JJ Lehto and Mika Salo. Finland has also produced most of the world's best rally drivers, including the ex-WRC World Champion drivers Marcus Gronholm, Juha Kankkunen, Hannu Mikkola, Tommi Makinen, Timo Salonen and Ari Vatanen. The only Finn to have won a road racing World Championship, Jarno Saarinen, was killed in 1973 while racing.

Among winter sports, Finland has been the most successful country in ski jumping, with former ski jumper Matti Nykanen being arguably the best ever in that sport.

Most notably, he won five Olympic medals (four gold) and nine World Championships medals (five gold). Among currently active Finnish ski jumpers, Janne Ahonen has been the most successful. Kalle Palander is a well-known alpine skiing winner, who won the World Championship and Crystal Ball (twice, in Kitzbuhel). Tanja Poutiainen has won an Olympic silver medal for alpine skiing, as well as multiple FIS World Cup races.

Some of the most outstanding athletes from the past include Hannes Kolehmainen (1890–1966), Paavo Nurmi (1897–1973) and Ville Ritola (1896–1982) who won eighteen gold and seven silver Olympic medals in the 1910s and 1920s.

They are also considered to be the first of a generation of great Finnish middle and long-distance runners (and subsequently, other great Finnish sportsmen) often named the "Flying Finns". Another long-distance runner, Lasse Viren (born 1949), won a total of four gold medals during the 1972 and 1976 Summer Olympics.

Riku Kiri, Jouko Ahola and Janne Virtanen have been the greatest strength athletes in the country, participating in the World's Strongest Man competition between 1993 and 2000.

The 1952 Summer Olympics, officially known as the *Games of the XV Olympiad*, were held in Helsinki, Finland. Other notable sporting events held in Finland include the 1983 and 2005 World Championships in Athletics, among others.

Some of the most popular recreational sports and activities include floorball, Nordic walking, running, cycling and skiing.

10

The Low Countries: Busy Lands

The Low Countries, in Dutch De Nederlanden, are the historical lands around the low-lying delta of the Rhine, Scheldt, and Meusc rivers, including the modern countries of Belgium, the Netherlands, Luxembourg and parts of northern France and western Germany. The term is more appropriate to the era of the Late Middle Ages and Early Modern Europe when strong centrally governed nations were slowly forming and territorial governance was in the hands of a noble or of a noble house.

Historically the region has its origins in Middle Francia, more precisely its northern part which became the Duchy of Lower Lotharingia. After the disintegration of Lower Lotharingia the Low Countries were brought under the rule of various stronger neighbours. Their possessions can be renamed into the Burgundian Netherlands and their succeeding Habsburg Netherlands, also called the United Seventeen Provinces (up to 1581), and later for the Southern parts as the Spanish Netherlands and Austrian Netherlands, whereas the northern parts formed the autonomous Dutch Republic. At times they reached a form of unity as the United Seventeen Provinces in the 16th Century, and later the United Kingdom of the Netherlands in the 19th Century.

Geo-political Situation

The term is not particularly current in modern contexts because the region does not exactly correspond to the sovereign states of the Netherlands, Belgium and Luxembourg, for which an alternative term, Benelux, was employed after the Second World War, but only to describe them as a trading union.

Before early modern nation building, the Low Countries referred to a wide area of northern Europe as a triangle roughly stretching from French Gravelines and Dunkirk at its southwestern point, to the area of Dutch Delfzijl and German Eastern Frisia at its northeastern point, and to Luxembourg and French Thionville in the southeast.

The Low Countries were the scene of the early northern towns, newbuilt rather than developed from ancient centres, that mark the reawakening of Europe in the 12th century. In that period, they became one of the most densely populated regions of Europe, together with northern Italy.

A collection of several regions rather than one homogeneous region, all the low countries still shared a great number of similarities.

- Most were coastal regions bounded by the North Sea or the English Channel. The countries not having access to the sea linked themselves politically and economically to those that had access, so as to form one union of port and hinterland. A poetic description also calls the region *the Low Countries by the Sea*
- Most spoke Middle Dutch out of which later would evolve Dutch. However some regions, such as the Bishopric of Liege, the Romance Flanders (around Cambrai, Lille, Tournai), the French-speaking part of Brabant around Nivelles and, Namur, where French or Walloon was the dominant language are often considered as part of the Low Countries as well.
- Most of them depended on a lord or count in name only, the cities effectively being ruled by guilds and councils and although in theory part of a kingdom, their interaction with their rulers was regulated by a strict set of liberties describing what the latter could and could not expect from them.
- All of them depended on trade and manufacturing and the encouragement of the free flow of goods and craftsmen.

Historical Situation

The low countries were part of the Roman provinces of Belgica, Germania Inferior and Germania Superior. They were inhabited by Belgic tribes, before these were replaced by Germanic tribes in the 4th and 5th century. They were governed by the ruling Merovingian dynasty.

By the end of the 8th century, the Low Countries formed a part of Francia and the Merovingians were replaced by the Carolingian dynasty. In 800 the Pope crowned and appointed Charlemagne Emperor of the re-established Roman Empire.

After the death of Charlemagne, Francia was divided in three parts among his three grandsons. The Low Countries became part of Middle Francia, which was ruled by Lothair I. After the death of Lothair, the Low Countries were coveted by the rulers of both West Francia and East Francia. Each tried to swallow the region and to merge it with their spheres of influence.

Thus, the Low Countries consisted of fiefs whose sovereignty resided with either the Kingdom of France or the Holy Roman Empire. The further history of the Low Countries can be seen as a continual struggle between these two powers.

Gradually, separate fiefs came to be ruled by a single family through intermarriage. This process culminated in the rule of the House of Valois, who were the rulers of the Duchy of Burgundy

In 1477 the Burgundian holdings in the area, the Burgundian Netherlands passed through an heiress — Mary of Burgundy — to the Habsburgs. In the following century the "Low Countries" corresponded roughly to the Seventeen Provinces covered by the Pragmatic Sanction of 1549 of Holy Roman Emperor Charles V, which freed the provinces from their archaic feudal obligations.

After the northern Seven United Provinces of the seventeen declared their independence from Habsburg Spain, the provinces of the Southern Netherlands were recaptured (1581) and are sometimes called the *Spanish Netherlands*.

In 1713, under the Treaty of Utrecht following the War of the Spanish Succession, what was left of the Spanish Netherlands was ceded to Austria and thus became known as the Austrian Netherlands. The United Kingdom of the Netherlands (1815-1830) temporarily united the Low Countries again.

Linguistic Distinction

In English, the plural form Netherlands is used for the present-day country, but in Dutch that plural has been dropped; one can thus distinguish between the older, larger Netherlands and the current country. So *Nederland* (singular) is used for the modern nation and *de Nederlanden* (plural) for the domains of Charles V. However: the official name of the Dutch kingdom is still Kingdom of the Netherlands (Koninkrijk der Nederlanden), in official use, the plural has not been dropped. The name Kingdom of the Netherlands also refers to the united kingdom of 1815-1830/39, which included present-day Belgium.

Netherlands

The Netherlands is a constituent country of the Kingdom of the Netherlands, located in North-West Europe. It is a parliamentary democratic constitutional monarchy. The Netherlands borders the North Sea to the north and west, Belgium to the south, and Germany to the east, and water borders with Denmark, Norway and United Kingdom. The capital is Amsterdam and the seat of government is The Hague.

The Netherlands in its entirety is often referred to as Holland, although North and South Holland are actually only two of its twelve provinces. The word *Dutch* is used to refer to the people, the language, and anything pertaining to the Netherlands. This lexical difference between the noun and the adjective is an attribute of the English language that does not exist in the Dutch language. The adjective 'Dutch' is derived from the language that was spoken in the area, called 'Diets', which equals Middle Dutch.

The Netherlands was one of the first parliamentary democracies. Among other affiliations the country is a founding member of the European Union (EU), NATO, OECD and WTO. With Belgium and Luxembourg it forms the Benelux economic union. The country is host to five international courts: the Permanent Court of Arbitration, the International Court of Justice, the International Criminal Tribunal for the Former

Yugoslavia, the International Criminal Court and the Special Tribunal for Lebanon. The first four are situated in The Hague as is the EU's criminal intelligence agency Europol and judicial co-operation agency Eurojust. This has led to the city being dubbed "the world's legal capital". The Netherlands has a capitalist market-based economy, ranking 15th of 157 countries according to the Index of Economic Freedom.

The Netherlands is a geographically low-lying country, with about 20% of its area and 21% of its population located below sea level, with 50% of its land lying less than one metre above sea level. Significant land area has been gained through land reclamation and preserved through an elaborate system of polders and dikes. Much of the Netherlands is formed by the estuary of three important European rivers, which together with their distributers form the Rhine-Meuse-Scheldt delta. Most of the country is very flat, with the exception of foothills in the far southeast and several low-hill ranges in the central parts.

History

Under Charles V, Holy Roman Emperor, and King of Spain, the region was part of the Seventeen Provinces of the Netherlands, which also included most of present-day Belgium, Luxembourg, and some land of France and Germany. The Eighty Years' War between the provinces and Spain began in 1568.

In 1579, the northern half of the Seventeen Provinces formed the Union of Utrecht, a treaty in which they promised to support each other in their defense against the Spanish army. The Union of Utrecht is seen as the foundation of the modern Netherlands. In 1581 the northern provinces adopted the Act of Abjuration, the declaration of independence in which the provinces officially deposed Philip II of Spain.

Queen Elizabeth I of England sympathized with the Dutch struggle against the Spanish, and in 1585 she concluded a treaty with the Dutch whereby she promised to send an English army to the Netherlands to aid the Dutch in their war with the Spanish. In December 1585, 7,600 soldiers were sent to the Netherlands from England under the command of Robert Dudley, 1st Earl of Leicester. However, the English army was wasted away in fruitless military campaigns in the Netherlands and had no real effect of helping the Dutch rebellion.

Robert Dudley returned to the Netherlands in November 1586 with another army. However, the result was no better than it had been in 1585. Philip II, the son of Charles V, was not prepared to let them go easily, and war continued until 1648, when Spain under King Philip IV finally recognised the independence of the seven northwestern provinces in the Peace of Munster. Parts of the southern provinces became *de facto* colonies of the new republican-mercantile empire.

Dutch Republic 1581–1795

After independence, the provinces of Holland, Zeeland, Groningen, Friesland, Utrecht, Overijssel, and Gelre formed a confederation known as the Republic of the Seven United Netherlands. All these provinces were autonomous and had their own government, the "States of the Province". The States-General, the confederal government, were seated in The Hague and consisted of representatives from each of the seven provinces. The sparsely populated region of Drenthe, mainly consisting of poor peatland, was part of the republic too, although Drenthe was not considered one of the provinces.

Drenthe had its own states, but the landdrost of Drenthe was appointed by the States-General. The Republic occupied a number of so-called Generality Lands (*Generaliteitslanden* in Dutch). These territories were governed directly by the States-General, so they did not have a government of their own and they did not have representatives in the States-General. Most of these territories were occupied during the Eighty Years' War. They were mainly Roman Catholic and were used as a buffer zone between the Republic and the Southern Netherlands.

The Dutch Empire grew to become one of the major seafaring and economic powers of the 17th century. In the Dutch Golden Age ("Gouden Eeuw"), colonies and trading posts were established all over the world. Dutch settlement in North America began with the founding of New Amsterdam, on the southern tip of Manhattan in 1614. In South Africa, the Dutch settled the Cape Colony in 1652. By 1650, the Dutch owned 16,000 merchant ships. During the 17th century, the Dutch population increased from an estimated 1.5 million to almost

2 million. Many economic historians regard the Netherlands as the first thoroughly capitalist country in the world. In early modern Europe it featured the wealthiest trading city (Amsterdam) and the first full-time stock exchange. The inventiveness of the traders led to insurance and retirement funds as well as such less benign phenomena as the boom-bust cycle, the world's first asset-inflation bubble, the tulip mania of 1636–1637, and, according to Murray Sayle, the world's first bear raider, Isaac le Maire, who forced prices down by dumping stock and then buying it back at a discount. The republic went into a state of general decline in the later 18th century, with economic competition from England and long standing rivalries between the two main factions in Dutch society, the *Staatsgezinden* (Republicans) and the *Prinsgezinden* (Royalists or Orangists) as main factors.

In the 17th century, plantation colonies were established by the Dutch and English along the many rivers in the fertile Guyana plains. The earliest documented colony in Guiana was along the Suriname River and called Marshall's Creek. The area was named after an Englishman. Disputes arose between the Dutch and the English. In 1667, the Dutch decided to keep the nascent plantation colony of Suriname conquered from the English, resulting from the Treaty of Breda. The English were left with New Amsterdam, a small trading post in North America, which is now known as New York City.

French Domination (1795–1814)

On 19 January 1795, one day after stadtholder William V of Orange fled to England, the *Bataafse Republiek* (Batavian Republic) was proclaimed, rendering the Netherlands a unitary state. From 1795 to 1806, the Batavian Republic designated the Netherlands as a republic modelled after the French Republic.

From 1806 to 1810, the *Koninkrijk Holland* (Kingdom of Holland) was set up by Napoleon Bonaparte as a puppet kingdom governed by his brother Louis Bonaparte in order to control the Netherlands more effectively. The name of the leading province, Holland, was used for the whole country. The Kingdom of Holland covered the area of the present day Netherlands, with the exception of Limburg and parts of Zeeland, which were French territory. In 1807, Prussian East Frisia and Jever were added to the kingdom. In 1809, however, after a failed British invasion, Holland had to give over all territories south of the Rhine to France.

King Louis Bonaparte did not meet Napoleon's expectations— he tried to serve Dutch interests instead of his brother's — and he was forced to abdicate on 1 July 1810. He was succeeded by his five-year-old son Napoleon Louis Bonaparte. Napoleon Louis reigned as Louis II for just ten days as Napoleon ignored his young nephew's accession to the throne. The Emperor sent in an army to invade the country and dissolved the Kingdom of Holland. The Netherlands then became part of the French Empire.

The Netherlands remained part of the French Empire until the autumn of 1813, when Napoleon was defeated in the battle of Leipzig and forced to withdraw his troops from the country.

Kingdom of the Netherlands

William I of the Netherlands, son of the last stadtholder William V van Oranje, returned to the Netherlands in 1813 and became Sovereign Prince of the Netherlands. On 16 March 1815, the Sovereign Prince became King of the Netherlands.

In 1815, the Congress of Vienna formed the United Kingdom of the Netherlands by expanding the Netherlands with Belgium in order to create a strong country on the northern border of France. In addition, William became hereditary Grand Duke of Luxembourg. The Congress of Vienna gave Luxembourg to William as personal property in exchange for his German possessions, Nassau-Dillenburg, Siegen, Hadamar, and Diez.

Belgium rebelled and gained independence in 1830, while the personal union between Luxembourg and the Netherlands was severed in 1890, when King William III of the Netherlands died with no surviving male heirs. Ascendancy laws prevented his daughter Queen Wilhelmina from becoming the next Grand Duchess. Therefore the throne of Luxembourg passed over from the House of Orange-Nassau to the House of Nassau-Weilburg, a junior branch of the House of Nassau.

The largest Dutch settlement abroad was the Cape Colony. It was established by Jan van Riebeeck on behalf of the Dutch East India Company at Cape Town (Dutch: *Kaapstad*) in 1652. The Prince of Orange acquiesced to British occupation and control of the Cape Colony in 1788. The Netherlands also possessed several other colonies, but Dutch settlement in these lands was limited. Most notable were the vast Dutch East Indies (now Indonesia) and Dutch Guiana (now Suriname). These 'colonies' were first administered by the Dutch East India Company and the Dutch West India Company, both collective private enterprises. Three centuries later these companies got into financial trouble, and the territories in which they operated were taken over by the Dutch government (in 1815 and 1791 respectively). Only then did they become official colonies.

During its colonial period the Netherlands was heavily involved in the slave trade. The Dutch planters relied heavily on African slaves to cultivate the coffee, cocoa, sugar cane and cotton plantations along the rivers. Treatment of the slaves by their owners was notoriously bad, and many slaves escaped the plantations.

Slavery was abolished by the Netherlands in Dutch Guiana in 1863, but the slaves were not fully released until 1873, after a mandatory 10 year transition period during which time they were required to work on the plantations for minimal pay and without state sanctioned torture. As soon as they became truly free, the slaves largely abandoned the plantations where they had suffered for several generations in favour of the city Paramaribo. Every year this is remembered during Keti Koti, 1 July, Emancipation Day (end of slavery).

During the 19th century, the Netherlands was slow to industrialize compared to neighbouring countries, mainly because of the great complexity involved in modernizing the infrastructure, consisting largely of waterways, and the great reliance its industry had on windpower.

Although the Netherlands remained neutral during World War I, it was heavily involved in the war. German general Count Schlieffen, who was Chief of the Imperial German General Staff had originally planned to invade the Netherlands while advancing into France in the original Schlieffen Plan. This was changed by Schlieffen's successor Helmuth von Moltke the Younger in order to maintain Dutch neutrality. Later during the war Dutch neutrality proved essential to German survival until the blockade by Great Britain in 1916, when the import of goods through the Netherlands was no longer possible.

However, the Dutch were able to continue to remain neutral during the war using their diplomacy and their ability to trade.

Second World War

The Netherlands intended to remain neutral during the Second World War. There were, however, contingency plans involving the armies of Belgium, France and the United Kingdom. Regardless, Nazi Germany invaded the Netherlands on 10 May 1940 as part of their campaign against the Allied forces. French forces in the south and British ships in the west came to help but turned around quickly, evacuating many civilians and several thousand German prisoners of war from the German elite airborne divisions.

The country was overrun in five days. Only after, but not because of, the bombing of Rotterdam the main element of the Dutch army surrendered on 14 May 1940; although a Dutch and French force held the western part of Zeeland for some time after the surrender. The Kingdom as such, continued the war from the colonial empire; the government in exile resided in London.

During the occupation, over 100,000 Dutch Jews were rounded up to be transported to Nazi German concentration camps in Germany, German-occupied Poland and German-occupied Czechoslovakia. By the time these camps were liberated, only 876 Dutch Jews survived. Dutch workers were conscripted for forced labour in German factories, civilians were killed in reprisal for attacks on German soldiers, and the countryside was plundered for food for German soldiers in the Netherlands and for shipment to Germany. Although there were thousands of Dutch who risked their lives by hiding Jews from the Germans, as recounted in The Hiding Place (book) by Corrie ten Boom and *The Heart Has Reasons* by Mark Klempner, there were also thousands of Dutch who collaborated with the occupying force in hunting down hiding Jews. Local fascists and anti-Bolsheviks joined the Waffen-SS in the 4th SS Volunteer Panzergrenadier Brigade Netherlands, fighting on the Eastern Front as well as other units.

On 8 December 1941, the Netherlands declared war on Japan. The government-in-exile then lost control of its major colonial stronghold, the Netherlands East Indies (Indonesia), to Japanese forces in March 1942. "American-British-Dutch-Australian" (ABDA) forces fought hard in some instances but were overwhelmed. During the Japanese occupation of Indonesia, the Japanese interned Dutch civilians and used Dutch and Indos (Eurasians of Dutch and Indonesian descent) alike as forced labour, both in the Netherlands East Indies and in neighbouring countries.

This included forcing women to work as "comfort women" (sex slaves) for Japanese personnel. The Dutch Red Cross reported the deaths in Japanese custody of 14,800 European civilians out of 80,000 interned and 12,500 of the 34,000 POW captured.

A later U.N. report stated that 4 million people died in Indonesia as a result of famine and forced labour (known as romusha) during the Japanese occupation. Some military personnel escaped to Australia and other Allied countries from where they carried on the fight against Japan. The Japanese furthered the cause of independence for the colony, so that after VE day many young Dutchmen found themselves fighting a colonial war against the new republic of Indonesia.

Princess Juliana of the Netherlands, the only child of Queen Wilhelmina and heir to the throne, sought refuge in Ottawa, Canada, with her two daughters, Beatrix and Irene, during the war. During Princess Juliana's stay in Canada, preparations were made for the birth of her third child. To ensure the Dutch citizenship of this royal baby, the Canadian Parliament passed a special law declaring Princess Juliana's suite at the Ottawa Civic Hospital "extraterritorial".

On 19 January 1943, Princess Margriet was born. The day after Princess Margriet's birth, the Dutch flag was flown on the Peace Tower. This was the only time in history a foreign flag has waved atop Canada's Parliament Buildings. In 1944-45, the First Canadian Army was responsible for liberating much of the Netherlands from German occupation. The joyous "Canadian summer" that ensued after the liberation, forged deep and long-lasting bonds of friendship between the Netherlands and Canada. In 1949, Dutch troops occupied an area of 69 km (27 sq mi) of the British zone of occupied Germany and annexed it. At that time, these areas were inhabited by almost 10,000 people.

Recent History

After the war, the Dutch economy prospered by leaving behind an era of neutrality and gaining closer ties with neighbouring states. The Netherlands was one of the founding members of the Benelux (Belgium, the Netherlands and Luxembourg) grouping, was among the twelve founding members of the North Atlantic Treaty Organisation (NATO), and was among the six founding members of the European Coal and Steel Community, which would later evolve, via the EEC (Common Market), into the European Union.

The 1960s and 1970s were a time of great social and cultural change, such as rapid *ontzuiling* (literally: depillarisation), a term that describes the decay of the old divisions along class and religious lines. Youths, and students in particular, rejected traditional mores and pushed for change in matters like women's rights, sexuality, disarmament and environmental issues. Today, the Netherlands is regarded as a liberal country, considering its drugs policy and its legalisation of euthanasia. Same-sex marriage has been permitted since 1 April 2001.

Geography

Rivers

The country is divided into two main parts by three large rivers, the Rhine (*Rijn*) and its main distributers, the Waal and the Meuse (*Maas*). These rivers functioned as a natural barrier between earlier fiefdoms and hence created traditionally a cultural divide, as is evident in some phonetic traits that are recognizable north and south of these "Large Rivers" (*de Grote Rivieren*). The southwestern part of the Netherlands is a river delta and two tributaries of the Scheldt (*Westerschelde and Oosterschelde*). Only one significant branch of the Rhine flows northeastward, the IJssel river, discharging into the IJsselmeer, the former Zuiderzee ('southern sea'). This river also forms a linguistic divide: people to the east of this river speak Dutch Low Saxon dialects (except for the province of Friesland, which has its own language).

Floods

Over the centuries, the Dutch coastline has changed considerably as a result of human intervention and natural disasters. Most notable in terms of land loss is the 1134 storm, which created the archipelago of Zeeland in the southwest.

On 14 December 1287, St. Lucia's flood affected the Netherlands and Germany killing more than 50,000 people in one of the most destructive floods in recorded history. The St. Elizabeth flood of 1421 and the mismanagement in its aftermath destroyed a newly reclaimed polder, replacing it with the 72-square-kilometre (28 sq mi) *Biesbosch* tidal floodplains in the south-centre. The last major flood in the Netherlands took place in early February 1953, when a huge storm caused the collapse of several dikes in the southwest of the Netherlands. More than 1,800 people drowned in the ensuing inundations. The Dutch government subsequently decided on a large-scale program of public works (the "Delta Works") to protect the country against future flooding. The project took more than thirty years to complete.

The disasters were partially increased in severity through human influence. People had drained relatively high lying swampland to use it as farmland. This drainage caused the fertile peat to compress and the ground level to drop, whereby they would lower the water level to compensate for the drop in ground level, causing the underlying peat to compress even more. The problem remains unsolvable to this day. Also, up until the 19th century peat was mined, dried, and used for fuel, further adding to the problem.

To guard against floods, a series of defences against the water were contrived. In the first millennium AD, villages and farmhouses were built on man-made hills called *terps*. Later, these terps were connected by dikes. In the 12th century, local government agencies called *"waterschappen"* (English "water bodies") or *"hoogheemraadschappen"* ("high home councils") started to appear, whose job it was to maintain the water level and to protect a region from floods. (These agencies exist to this day, performing the same function.) As the ground level dropped, the dikes by necessity grew and merged into an integrated system. By the 13th century, windmills had come into use in order to pump water out of areas below sea level. The windmills were later used to drain lakes, creating the famous polders.

In 1932, the *Afsluitdijk* (English "Closure Dike") was completed, blocking the former *Zuiderzee* (Southern Sea) from the North Sea and thus creating the IJsselmeer (IJssel Lake). It became part of the larger Zuiderzee Works in which four polders totalling 2,500 square kilometres (965 sq mi) were reclaimed from the sea.

Delta Works

After the 1953 disaster, the Delta project, a vast construction effort designed to end the threat from the sea once and for all, was launched in 1958 and largely completed in 1997 with the completion of the Maeslantkering. The official goal of the Delta project was to reduce the risk of flooding in South Holland and Zeeland to once per 10,000 years. (For the rest of the country, the protection level is once per 4,000 years.) This was achieved by raising 3,000 kilometers (1,864 mi) of outer sea-dykes and 10,000 kilometers (6,214 mi) of inner, canal, and river dikes to "delta" height, and by closing off the sea estuaries of the Zeeland province. New risk assessments occasionally show problems requiring additional Delta project dyke reinforcements. The Delta project is one of the largest construction efforts in human history and is considered by the American Society of Civil Engineers as one of the seven wonders of the modern world.

Additionally, the Netherlands is one of the countries that may suffer most from climate change. Not only is the rising sea a problem, but also erratic weather patterns may cause the rivers to overflow.

Climate

The predominant wind direction in the Netherlands is southwest, which causes a moderate maritime climate, with cool summers and mild winters.

Nature

The Netherlands has 20 national parks and hundreds of other nature reserves. Most are owned by Staatsbosbeheer and Natuurmonumenten and include lakes, heathland, woods, dunes and other habitats.

Phytogeographically, the Netherlands is shared between the Atlantic European and Central European provinces of the Circumboreal Region within the Boreal Kingdom. According to the World Wide Fund for Nature, the territory of the Netherlands belongs to the ecoregion of Atlantic mixed forests.

In 1871 the last old original natural woods (Beekbergerwoud) were cut down, and most woods today are planted monocultures of trees like Scots Pine and trees that are not native to the Netherlands. These woods were planted on anthropogenic heaths and sand-drifts (overgrazed heaths) (Veluwe).

Economy

The Netherlands has a prosperous and open economy in which the government has reduced its role since the 1980s. Industrial activity is predominantly in food-processing (Unilever, Heineken), financial services (ING), chemicals (DSM), petroleum refining (Shell), and electrical machinery (Philips, ASML).

The Netherlands has the 16th largest economy in the world, and ranks 10th in GDP (nominal) per capita. Between 1998 and 2000 annual economic growth (GDP) averaged nearly 4%, well above the European average. Growth slowed considerably from 2001 to 2005 with the global economic slowdown, but accelerated to 4.1% in the third quarter of 2007. Inflation is 1.3%, and unemployment is at 4.0% of the labour force. By Eurostat standards, unemployment in the Netherlands is at 4.1% (April 2010)-the lowest rate of all European Union member states. The Netherlands also has a relatively low GINI coefficient of 0.326. Despite ranking only 10th in GDP per capita, UNICEF ranked the Netherlands 1st in child well-being. On the Index of Economic Freedom Netherlands is the 13th most free market capitalist economy out of 157 surveyed countries.

The Netherlands introduced the euro in 1999. It is one of the 16 sovereign states that make up the Eurozone.

Amsterdam is the financial and business capital of the Netherlands. The Amsterdam Stock Exchange (AEX), part of Euronext, is the world's oldest stock exchange and is one of Europe's largest bourses. It is situated near Dam Square in the city's centre. As a founding member of the euro, the Netherlands replaced (for accounting purposes) its former currency, the "Gulden" (guilder), on 1 January 1999, along with 15 other adopters of the Euro. Actual euro coins and banknotes followed on 1 January 2002. One euro was equivalent to 2.20371 Dutch guilders.

The Netherlands' location gives it prime access to markets in the UK and Germany, with the port of Rotterdam being the largest port in Europe. Other important parts of the economy are international trade (Dutch colonialism started with cooperative private enterprises such as the VOC), banking and transport. The Netherlands successfully addressed the issue of public finances and stagnating job growth long before its European partners. Amsterdam is the 5th busiest tourist destination in Europe with more than 4.2 million international visitors.

The country continues to be one of the leading European nations for attracting foreign direct investment and is one of the five largest investors in the U.S. The economy experienced a slowdown in 2005, but in 2006 recovered to the fastest pace in six years on the back of increased exports and strong investment. The pace of job growth reached 10-year highs in 2008. The Netherlands moved up from the 11th position in the Global Competitiveness Index to the 9th position in 2007

Infrastructure, Agriculture and Natural Resources

A Frisian Holstein cow in the Netherlands: Intensive dairy farming is an important part of agriculture.

Rotterdam has the largest port in Europe, with the rivers Meuse and Rhine providing excellent access to the hinterland upstream reaching to Basel, Switzerland, and into France. In 2006, Rotterdam was the world's seventh largest container port in terms of Twenty-foot equivalent units (TEU) handled. The port's main activities are petrochemical industries and general cargo handling and transshipment. The harbour functions as an important transit point for bulk materials and between the European continent and overseas. From Rotterdam goods are transported by ship, river barge, train or road. In 2007, the Betuweroute, a new fast freight railway from Rotterdam to Germany, was completed.

A highly mechanised agricultural sector employs 4% of the labour force but provides large surpluses for the food-processing industry and for exports. The Dutch rank third worldwide in value of agricultural exports, behind the United States and France, with exports earning $55 billion annually. A significant portion of Dutch agricultural exports are derived from fresh-cut plants, flowers, and bulbs, with the Netherlands exporting two-thirds of the world's total. The Netherlands also exports a quarter of all world tomatoes, and one-third of the world's exports of chilis and cucumbers.

One of the largest natural gas fields in the world is situated near Slochteren. Exploitation of this field resulted in a total revenue of €159 billion since the mid 1970s. With just over half of the reserves used up and an expected continued rise in oil prices, the revenues over the next few decades are expected to be at least that much.

Government and Administration

Government

The Netherlands has been a constitutional monarchy since 1815 and a parliamentary democracy since 1848. The Netherlands is described as a consociational state. Dutch politics and governance are characterised by an effort to achieve broad consensus on important issues, within both the political community and society as a whole. In 2008, The Economist ranked The Netherlands as the fourth most democratic country in the world.

The monarch is the head of state, at present Queen Beatrix. Constitutionally, the position is equipped with limited powers. The monarch can exert some influence during the formation of a new cabinet, where they serve as neutral arbiter between the political parties. Additionally, the king (the title queen has no constitutional significance) has the right to be informed and consulted. Depending on the personality and qualities of the king and the ministers, the king might have influence beyond the power granted by the constitution.

In practice, the executive power is formed by the ministerraad, the deliberative council of the Dutch cabinet. The cabinet consists usually of thirteen to sixteen ministers and a varying number of state secretaries. One to three ministers are ministers without portfolio. The head of government is the Prime Minister of the Netherlands, who often is the leader of the largest party of the coalition. In fact, this has been continuously the case since 1973. The Prime Minister is a primus inter pares, meaning he has no explicit powers beyond those of the other ministers. Currently, the Prime Minister is Jan Peter Balkenende.

The cabinet is responsible to the bicameral parliament, the States-General which also has legislative powers. The 150 members of the House of Representatives, the Lower House, are elected in direct elections, which are held every four years or after the fall of the cabinet (by example: when one of the chambers carries a motion of no-confidence, the cabinet offers its resignation to the monarch). The States-Provincial are directly elected every four years as well. The members of the provincial assemblies elect the 75 members of the Senate, the upper house, which has less legislative powers, as it can merely reject laws, not propose or amend them.

Both trade unions and employers organisations are consulted beforehand in policymaking in the financial, economic and social areas. They meet regularly with government in the Social-Economic Council. This body advises government and its advice cannot be put aside easily.

While historically the Dutch foreign policy was characterised by neutrality, since the Second World War the Netherlands became a member of a large number of international organisations, most prominently the UN, NATO and the EU. The Dutch economy is very open and relies on international trade.

The Netherlands has a long tradition of social tolerance. In the 18th century, while the Dutch Reformed Church was the state religion, Catholicism and Judaism were tolerated. In the late 19th century this Dutch tradition of religious tolerance transformed into a system of pillarisation, in which religious groups coexisted separately and only interacted at the level of government. This tradition of tolerance is linked to Dutch criminal justice policies on recreational drugs, prostitution, LGBT rights, euthanasia, and abortion which are among the most liberal in the world.

Political Parties

Due to the multi-party system, no single party has held a majority in parliament since the 19th century, and coalition cabinets had to be formed. Since suffrage became universal in 1919, the Dutch political system has been dominated by three families of political parties: the strongest family were the Christian democrats, currently represented by the Christian

Democratic Appeal (CDA), second were the social democrats, of which the Labour Party (PvdA), and third were the liberals, of which the People's Party for Freedom and Democracy (VVD) is the main representative.

These parties cooperated in coalition cabinets in which the Christian democrats had always been a partner: so either a centre left coalition of the Christian democrats and social democrats was ruling or a centre right coalition of Christian democrats and liberals. In the 1970s, the party system became more volatile: the Christian democratic parties lost seats, while new parties became successful, such as the radical democrat and progressive liberal D66.

In the 1994 election, the CDA lost its dominant position. A "purple" cabinet was formed by VVD, D66, and PvdA. In the 2002 elections, this cabinet lost its majority, due to an increased support for the CDA and the rise of the LPF, a new political party around the flamboyant populist Pim Fortuyn, who was assassinated a week before the elections. A short-lived cabinet was formed by CDA, VVD, and LPF, which was led by the CDA leader, Jan Peter Balkenende. After the 2003 elections in which the LPF lost most of its seats, a cabinet was formed by CDA, VVD, and D66. The cabinet initiated an ambitious program of reforming the welfare state, the health care system, and the immigration policies.

In June 2006, the cabinet fell after D66 voted in favour of a motion of no confidence against minister of immigration and integration Rita Verdonk, who had instigated an investigation of the asylum procedure of VVD MP Ayaan Hirsi Ali. A care taker cabinet was formed by CDA and VVD, and the general elections were held on 22 November 2006. In these elections, the CDA remained the largest party and the SP made the largest gains. The formation of a new cabinet took three months, resulting in a coalition of CDA, PvdA, and Christian Union.

On 20 February 2010, the cabinet fell as the PvdA refused to prolong the involvement of the Dutch Army in Uruzgan, Afghanistan. Snap elections were held on 9 June 2010, with devastating results for the previously largest party, the CDA, which lost about half of its seats, resulting in 21 seats. The VVD became the largest party with 31 seats, closely followed by the PvdA with 30 seats. The big winner of the 2010 elections was Geert Wilders' PVV which more than doubled in number of seats. Negotiation talks for a new government have started, with the potential outcome still being uncertain.

Administrative Divisions

The Netherlands is divided into twelve administrative regions, called provinces, each under a Governor, who is called Commissaris van de Koningin (Commissioner of the Queen), except for the province Limburg where the commissioner is called Gouverneur (Governor). All provinces are divided into municipalities (gemeenten), 431 in total (1 January 2010).

The country is also subdivided in water districts, governed by a water board (waterschap or hoogheemraadschap), each having authority in matters concerning water management. As of 1 January 2005 there are 27. The creation of water boards actually pre-dates that of the nation itself, the first appearing in 1196. In fact, the Dutch water boards are one of the oldest democratic entities in the world still in existence.

Demographics

The Netherlands has an estimated population of 16,491,852 (as of 8 March 2009). It is the 11th most populous country in Europe and the 61st most populous country in the world. Between 1900 and 1950, the country's population almost doubled from 5.1 to 10.0 million people. From 1950 to 2000, the population further increased from 10.0 to 15.9 million people, but the population growth decreased compared to the previous fifty years. The estimated growth rate is currently 0.436% (as of 2008).

The fertility rate in the Netherlands is 1.66 children per woman (as of 2008), which is high compared to many other European countries, but well below the 2.1-rate required for natural population replacement. Life expectancy is high in the Netherlands: 82 years for newborn girls and 77 for boys (2007). The country has a migration rate of 2.55 migrants per 1,000 inhabitants.

The majority of the population of the Netherlands are ethnically Dutch. A 2005 estimate counted: 80.9% Dutch, 2.4% Indonesian (Indo-Dutch, South Moluccan), 2.4% German, 2.2% Turkish, 2.0% Surinamese, 1.9% Moroccan, 0.8% Antillean and

Aruban, and 6.0% others. The Dutch people are among the tallest in the world, with an average height of about 1.81 m (5 ft 11 in) for adult males and 1.68 m (5 ft 6 in) for adult females. People in the south are on average about 2 cm shorter than those in the north.

The Netherlands is the 25th most densely populated country in the world, with 395 inhabitants per square kilometre (1,023 sq mi)—or 484 people per square kilometre (1,254/sq mi) if only the land area is counted. It is the most densely populated country in Europe with a population over 16 million. The Randstad is the country's largest conurbation located in the west of the country and contains the four largest cities: Amsterdam in the province North Holland, Rotterdam and The Hague in the province South Holland, and Utrecht in the province Utrecht. The Randstad has a population of 7 million inhabitants and is the 6th largest metropolitan area in Europe.

Dutch people, or descendants of Dutch people, are also found in migrant communities worldwide, notably in Canada, Australia, South Africa and the United States. According to the 2006 U.S. Census, more than 5 million Americans claim total or partial Dutch ancestry. There are close to 3 million Dutch-descended Afrikaners living in South Africa. In 1940, there were 290,000 Europeans and Eurasians in Indonesia, but most have since left the country.

The official language is Dutch, which is spoken by the vast majority of the inhabitants.

Another official language is Frisian, which is spoken in the northern province of Friesland, called Fryslan in that language. Frisian is also spoken in some villages in the west of the province of Groningen. Frisian is co-official only in the province of Friesland, although with a few restrictions. Several dialects of Low Saxon (Nedersaksisch in Dutch) are spoken in much of the north and east, like the Tweants in the Twente region, and are recognised by the Netherlands as regional languages according to the European Charter for Regional or Minority Languages, as well as the Meuse-Rhenish Franconian varieties in the southeastern province of Limburg, here called Limburgish language.

There is a tradition of learning foreign languages in the Netherlands: about 70% of the total population have good knowledge of conversational English, 55– 59% of German, and 19% of French. English is a mandatory course in all secondary schools. In most lower level secondary school educations (VMBO), one additional modern foreign language is mandatory during the first two years. In higher level secondary schools, two additional modern foreign languages are mandatory during the first three (HAVO) or four (VWO) years. The standard modern languages are French and German, although schools can change one of these modern languages with Spanish, Turkish, Arabic, or Russian. Additionally, schools in the Frisia region teach and have exams in Frisian, and schools across the country teach and have exams in ancient Greek and Latin.

Religion

The Netherlands is one of the most secular countries in Western Europe, with only 39% being religiously affiliated (31% for those aged under 35), and fewer than 20% visiting church regularly.

Currently, Roman Catholicism is the single largest religion of the Netherlands, forming the religious home of some 26.3% of the Dutch people, down from 40% in the 1970s. The Protestant Church of the Netherlands follows with 11.4% of the population. It was formed in 2004 as a merger of the two major strands of Calvinism: the Dutch Reformed Church (which represented roughly 8.5% of the population), the Reformed Churches in the Netherlands (3.7% of the population), and a smaller Lutheran Church. Other Protestant churches, mostly orthodox Calvinist splits, represent 6% of the population.

In 2006, there were 850,000 Muslims (5% of the total Dutch population). The Netherlands has an estimated 250,000 Buddhists or people who feel strongly attracted by this religion, largely white Dutch. There are approximately 95,000 Hindus, of whom 85% originate from Suriname. Sikhs are another religious minority numbering around 12,000, mainly located in or around Amsterdam. There are five gurudwaras in the Netherlands. The Association of Religion Data Archives (relying on World Christian Encyclopedia) estimated some 6,400 Bahá'ís in 2005.

Although the Holocaust deeply affected the Jewish community (killing about 75% of its 140,000 members at the time), it has managed to rebuild a vibrant and lively Jewish life for its approximately 45,000 current members. Before World War II, 10% of the Amsterdam population was Jewish.

According to the most recent Eurobarometer poll 2005, 34% of the Dutch citizens responded that "they believe there is a god", whereas 37% answered that "they believe there is some sort of spirit or life force", and 27% that "they do not believe there is any sort of spirit, god, or life force".

In 1947, before major secularisations in Europe and the large settlement of non-Europeans in the Netherlands, most Dutch citizens identified themselves as Christians. In the same year, 44.3% belonged to Protestant denominations, 38.7% belonged to the Roman Catholic Church, and 17.1% were unaffiliated. Since then, the general collapse in religiosity has struck Protestants somewhat harder than Catholics, which partly explains why the Catholic Church has a larger percentage than the Protestants nowadays. In 2003, those percentages were approximately: Protestant 20, Catholic 30 and unaffiliated 41.

Freedom of education has been guaranteed by the Dutch constitution since 1917, and schools run by religious groups (especially Christian and Islam) are funded by the government. Nevertheless, all schools must meet strict quality criteria.

Three political parties in the Dutch parliament (CDA, Christian Union, and SGP) base their policy on Christian belief in varying degrees. Although the Netherlands is a secular state, in some municipalities where the Christian parties have the majority, the council practices religion by praying before a meeting. Moreover, in a few remaining (rural) spots, roads are closed for car traffic on Sundays and religious holidays.

Municipalities in general also give civil servants a day off on Christian religious holidays, such as Easter and the Ascension of Jesus. On 4 September 2008, a discussion was started by Tineke Huizinga whether Islam should receive a holiday, since there are multiple for Christianity. In 2005, 20% of the Dutch thought that such a holiday should be a national holiday (which means the entire country receives a day off work or school) and 45% thought that Eid ul-Fitr should at least be recognized as a holiday.

Culture

The Netherlands has had many well-known painters. The 17th century, when the Dutch republic was prosperous, was the age of the "Dutch Masters", such as Rembrandt van Rijn, Johannes Vermeer, Jan Steen, Jacob van Ruysdael and many others. Famous Dutch painters of the 19th and 20th century were Vincent van Gogh and Piet Mondriaan. M. C. Escher is a well-known graphics artist. Willem de Kooning was born and trained in Rotterdam, although he is considered to have reached acclaim as an American artist. The Netherlands is the country of philosophers Erasmus of Rotterdam and Spinoza. All of Descartes' major work was done in the Netherlands. The Dutch scientist Christiaan Huygens (1629–1695) discovered Saturn's moon Titan and invented the pendulum clock. Antonie van Leeuwenhoek was the first to observe and describe single-celled organisms with a microscope.

In the Dutch Golden Age, literature flourished as well, with Joost van den Vondel and P.C. Hooft as the two most famous writers. In the 19th century, Multatuli wrote about the poor treatment of the natives in Dutch colonies. Important 20th century authors include Harry Mulisch, Jan Wolkers, Simon Vestdijk, Cees Nooteboom, Gerard (van het) Reve and Willem Frederik Hermans. Anne Frank's Diary of a Young Girl was published after she died in the Holocaust and translated from Dutch to all major languages.

Replicas of Dutch buildings can be found in Huis ten Bosch, Nagasaki, Japan. A similar Holland Village is being built in Shenyang, China. Windmills, tulips, wooden shoes, cheese, Delftware pottery, and cannabis are among the items associated with the Netherlands by tourists.

Education

The Netherlands has compulsory education from age 4 to 18 (or 16 as a study is completed which has given the student adequate professional skills to start as a professional in the labour market).

Pupils attend primary or elementary school from age 4 to age 12. After that they continue their education at secondary school minimally until age 18; which indicates one of three tracks in the Dutch educational system.

The vocational track starts with VMBO, which is seen as the lowest level of secondary education and lasts four years. Successfully completing VMBO results in a low level vocational degree and/or gives access to higher (secondary) levels vocational education. Completion of second level vocational education results in professional skills and gives access to further study a university of applied science.

The medium level HAVO lasts five years. After completion a student can attend a university of applied science, which award professional bachelor degrees. A degree at a university of applied science gives access to the university system.

The highest level of high school education is VWO, which lasts six years, completion of which allows students to attend a university. University consists of a three year bachelor's degrees, followed by one or two year master's degrees. A master's degree is required to start a four year doctoral degree. Doctoral candidates in the Netherlands are temporary employees of a university.

Military

The Netherlands has the oldest standing army in Europe; it was first established as such by Maurice of Nassau. The Dutch army was used throughout the Dutch Empire. After the defeat of Napoleon, the Dutch army was transformed into a conscription army. The army was unsuccessfully deployed during the Belgian revolution in 1830. After 1830, it was deployed mainly in the Dutch colonies, as the Netherlands remained neutral in European wars (including WWI), until the Netherlands was invaded in WWII and quickly conquered by the Wehrmacht in May 1940.

After WWII, the Netherlands dropped their neutrality, and the Dutch army became part of the NATO army strength in Cold War Europe; holding several bases in Germany. In 1996 conscription was suspended, and the Dutch army was once again transformed into a professional army. Since the 1990s the Dutch army has been involved in the Bosnian War, the Kosovo War, has been holding a province in Iraq after the defeat of Saddam Hussein, and is currently engaged in Afghanistan. The military is composed of four branches, all of which carry the prefix *Koninklijke* (Royal):

- *Koninklijke Landmacht* (KL), the Royal Netherlands Army
- *Koninklijke Marine* (KM), the Royal Netherlands Navy, including the Naval Air Service and Marine Corps
- *Koninklijke Luchtmacht* (KLu), the Royal Netherlands Air Force
- *Koninklijke Marechaussee* (KMar), the Royal Military Police, tasks include military police and border control.

General Peter van Uhm is the current Commander of the Netherlands armed forces. All military specialities except the submarine service and the Royal Netherlands Marine Corps (*Korps Mariniers*) are open to women. The Korps Commandotroepen, the Special Operations Force of the Netherlands Army, is open to women, but because of the extremely high physical demands for initial training, it is found impossible for women to become a commando. The Dutch Ministry of Defence employs almost 70,000 personnel, including over 20,000 civilian and over 50,000 military personnel.

Bibliography

Belden C.: *Landscapes of the Sacred: Geography and Narrative in American Spirituality*, New York: Paulist Press, 1988.

Berry, J.K. : *Beyond Mapping, Concepts, Algorithms and Issues in GIS*, Fort Collins, CO: GIS World Books, 1993.

Bolstad, P. : *GIS Fundamentals, A first text on Geographic Information Systems, Second Edition*, White Bear Lake, MN, Eider Press, 2005.

Cronon, William: *Changes in the Land: Indians, Colonists, and the Ecology of New England.* New York: Hill and Wang, 1983.

Davis, Mike: *Ecology of Fear: Los Angeles and the Imagination of Disaster*. New York: Vintage, 1999.

Elangovan K. : *GIS, Fundamentals, Applications and Population Geography*, New India Publishing Agency, New Delhi 2006.

Harvey, David: *Explanation in Geography.* London: Edward Arnold, 1969.

Harvey, Francis : *A Primer of GIS, Fundamental Geographic and Cartographic Concepts,* The Guilford Press, 2008.

Jackson, Peter: *Maps of Meaning: An Introduction to Cultural Geography.* London and Boston: Unwin Hyman, 1989.

James S. Duncan: *Writing Worlds: Discourse, Text, and Metaphor in the Representation of Landscape.* London and New York: Routledge, 1992.

James, T.: *The Best Poor Man's Country: A Geographical Study of Early Southeastern Pennsylvania*. Baltimore: Johns Hopkins Press, 1972.

King, Anthony D.: *Buildings and Society: Essays on the Social Development of the Built Enviroment*. London: Routledge & Kegan Paul, 1980.

Lewis, Peirce. *New Orleans: The Making of an Urban Landscape.* Cambridge, MA: Ballinger Publishing Co., 1976. Annotation

Liam Kennedy: *Urban Space and Representation.* Sterling, VA: Pluto Press, 2000.

Lung, Y., and van Tulder, R. : *Cars, Carriers of Regionalism*, Palgrave Macmillan, New York, 2004.

McDonnell, R.A. : *Principles of Geographical Information Systems*, Oxford University Press, Oxford, 1998.

Michael P.: *The Making of the American Landscape.* Boston: Unwin Hyman, 1990.

Michael, H.: *City Form and Natural Process: Towards a New Urban Vernacular.* London: Croom Helm, 1984.

Peter, G.: *The Slow Plague: A Geography of the AIDS Pandemic.* Oxford and Cambridge, MA: Blackwell Publishers, 1993.

Ray Allen: *The Far Western Frontier, 1830-1860.* The New American Nation Series. New York: Harper, 1956.

Richard SCASE : *Managing Creativity,* Open University Press, 2000.

Stephen, D.: *Fields of Vision: Landscape Imagery and National Identity in England and the United States.* Cambridge: Polity, 1993.

Tulder, R. : *Cars, Carriers of Regionalism*, Palgrave Macmillan, New York, 2004.

Victor, A. : *Geography of Art and Culture*, Contributions to Economic Analysis, Elsevier, 2004.

Watkins, Eric: *The Middle Eastern Environment*, London, The British Society for Middle Eastern Studie, 1995.

William G. BOWEN : *Performing Arts, The Economic Dilemma,* Thousand Oaks, SAGE, London, 2002.

Worboys, Michael, and Matt Duckham. : *GIS, a Computing Perspective*, Boca Raton, CRC Press, 2004.

Index

A

B

C

D

E

□□□